SONG OF THE SOUL

H.H. Roop Chandji Maharaj

Roop Chandji Maharaj was a great yogi and enlightened master
in the family of monks to which Guruji belongs. He left his
body 100 years ago. Maharajji is Guruji's spiritual guru.

Late Shri Devendra Jain
of Munshiram Manoharlal Jain family, Delhi
Smt Nirmal Jain has generously sponsored the publication of
this new edition in fond memory of her husband.

SONG OF THE SOUL

An Introduction to the Namokar Mantra
and the Science of Sound

H. H. Acharya Sushil Kumarji Maharaj

DEV BOOKS

Published for

Acharya Sushil Ashram

C-599, Chetna Marg
Defence Colony,
New Delhi 110 024
Tel: 24335560

by

DEV BOOKS

11-B Court Road
Civil Lines
Delhi 110 054
Tel: 9810236140
E-mail: devbooks@vsnl.net

ISBN 81-89835-07-6
This edition 2007
© Acharya Sushil Ashram, 2007

Printed in India by Dev Books, 11-B Court Road, Delhi 110 054

CONTENTS

Sound and the Serpent Power

PART II : GENERAL MANTRA KNOWLEDGE

PART III : TECHNIQUES FOR PRACTICE

About the Author

H. H. ACHARYA SUSHIL KUMARJI MAHARAJ is a self-realized master who has devoted more than fifty years to promoting peace, non-violence and knowledge of the self.

Guruji (as he is lovingly called by many of his devotees) was born 15 June 1926 in Sikhopur, a small foothill village in Haryana, India. The village was later named Sushilgarh in Guruji's honor. As a seven-year-old he left his home to live with Shri Chotelalji Maharaj, who later became his religious guru.

When Guruji was still a young boy, Shri Roop Chandji Maharaj appeared to him in spirit and told him to become a monk. (Maharaj was a great yogi and enlightened master in the family of monks to which Guruji belongs. He left his body 100 years ago. Roop Chandji Maharaj is Guruji's spiritual guru.) In this life, Guruji was not taught yogic systems from any master. His knowledge was realized through direct experience, and his powers were awakened through the grace of past lives. When he was fifteen years old he became a Jain muni (monk) in the Swetambar Sthanakvasi sect.

During his academic career in India, he passed through a number of examinations such as *Shastri, Acharya, Sahitya-Ratna, Vidya-Ratna, etc,* and mastered the classical studies of Indian religious and yogic philosophies.

Dedicated to religious harmony and the non-violent path, Guruji founded The World Fellowship of Religions and The International Mahavir Jain Mission in 1953 and 1977, respectively. Since his first controversial world tour in 1975 (for centuries Jain monks have travelled only on foot), he has travelled and taught extensively, spreading the message of non-violence and self-awareness. He has founded many ashrams and centers in the East and West. Guruji's main ashram and 108-acre retreat is Siddhachalam, located in Blairstown, New Jersey.

Guruji is a master of meditation and the science of sound. He is the originator of a teaching system which he called Arhum Yoga. It is an ancient system for the mastery of the inner self through watchfulness and direct perception. Arhum Yoga encompasses all aspects of philosophy and yogic practice in the Arihant spiritual tradition.

About Jainism

JAINISM IS A MODERN WORD to describe the ancient spiritual path of non-violence. The term 'Jain' comes from the word *jina* which means "spiritual victor". It designates a person who is a conqueror of the inner enemies – anger, greed, fear, attachment and hatred.

While there are about ten million adherents to the Jain religion in India today, and 250,000 abroad, Guruji teaches that Jainism is not simply a religion. It does not consist of just ceremony and ritual, but is, indeed, a science, a collection of techniques to know the Self. Many of the traditions and rituals which abide in this faith can be understood to possess a deep yogic significance.

Jainism is a way of life based on ahimsa or non-violence. Jain philosophy is ennobling and optimistic, where the soul of man and woman evolves to Godhood when matter no longer has any power over the soul. It embraces the ancient techniques and philosophies handed down by the *jinas,* who are also known as Arihantas or Tirthankaras. These are the enlightened preceptors in the Jain faith numbering twenty-four. This lineage can be traced back to prehistoric times. Lord Mahavir was the most recent Tirthankara (*circa* 500 BC), and Lord Parswanath was his predecessor.

2500 yrs. ago

Author's Preface

WHEN I WAS SEVEN YEARS OLD, I went to live with my Guruji, Shri Chotelalji Maharaj. I had been born into a Hindu Brahmin family, so it was from Guruji, a Jain monk, that I first learned the Namokar Mantra. The mantra immediately appealed to me. When I repeated it, it felt as if there were sparks of energy and light inside me.

At ten years old, in Jammu, Kashmir, I and my dear friend Sohan Lal Oswal decided to repeat the Namokar Mantra 125,000 times. We had not talked about it nor searched for any results or effects of the mantra; however, eight days after the completion of the japa, I realized the power of sound. During my stay in Jammu with Guruji, I regularly ate breakfast in a particular home. There was a girl there who was like a sister to me. One morning when I arrived, I found her crying and trembling – she had just been stung in the foot by a very poisonous black scorpion. People came and went, trying to help her by repeating various mantras, but nothing worked. I decided to try to remove the poison, applying the Namokar Mantra, even though I knew nothing about healing.

There was a dark blue line running from her toe to her thigh. I tied a handkerchief around her thigh and was thinking and affirming that the poison was going down, while directing the flow downward with my hand. It actually did begin to go down slowly, inch by inch, as I repeated the mantra. The girl began to feel some comfort as the poison was removed down to the ankle.

Then Guruji arrived, and I excitedly told him what had happened. I was very surprised and very happy that

I could feel the energy and see the mantra working, but despite this fact Gurüji ordered me not to continue. He said I should not do anything like that without perfect knowledge of the whole system.

The next day, despite the admonition from my guru, I returned to the girl, who had rested well that night, and saw that the poison was still only in her foot, and that the pain had considerably lessened. I repeated the mantra again that morning, and the poison completely left her body. This was the first time I experienced the power of mantra. From this point on my faith in the Namokar Mantra became unshakable.

A valuable question to ask yourself is – what is your goal? What do you want to become in this life? To this question, I must answer that the Namokar Mantra is my goal and my life. It is my love and my destiny. Through it I can serve and guide along the path of non-violence. We cannot limit the divine eternal sound of mantra by labelling it Jain, Buddhist, Christian, Hindu or Muslim. We should not label the Namokar Mantra under any classification just as we should not label ourselves.

The Namokar Mantra is for everyone. It is a song of the soul and used to attain realization of the perfection of the soul. The origin of sound and soul is the same, and both are eternal. It is both the source and the commentator of vast spiritual knowledge. From the non-violence of the ethical code or *Yama* to the bliss of self-realization or *Samadhi*, the Namokar Mantra guides us through the Eight Steps of Yoga.

Through its repetition, all qualities of evolved consciousness of the Arihanta, Siddha, Acharya, Upadhyaya and Sadhu are awakened. Right knowledge, Faith,

Conduct and Infinite Energy are attained. It holds the key to the science of sound, color and the universe. The Namokar Mantra is a great healer and can awaken the kundalini. It can purify and transform the self to the state of completeness. Japa or repetition of the mantra activates the many sleeping powers of the subconscious mind. It is similar to modern methods of auto-suggestion in this respect.

You repeat again and again, "I bow to the Arihant" [*Namo Arihantanam*]. You pay your respect, love and devotion to the Arihant. You humble yourself and give up your ego. You want to merge with the Arihant and Siddha. When you visualize them completely, all obstacles are removed, the 'inner enemies' are conquered and all divine powers are awakened.

In my youth and in many years which followed, wherever I used the Namokar Mantra, it proved immeasurably useful, whether for spiritual practice, healing or protection. Countless people have been healed of illnesses, have been protected in dangerous accidents and have experienced profound peace and deep meditation through the Namokar Mantra. So when I came to America, I decided to teach the science of sound vibration according to the Arihant tradition. The ancient teachings of the Arihantas are very powerful, very clear and true, and the Namokar Mantra is the essence of that knowledge. This small book will simply introduce you to the mantra's power. More detailed texts must follow in order to do justice to this vast subject.

I am happy that step-by-step, the people of the Siddhachalam ashram are preparing more teachings for publication. My blessings to Gurushakti for compiling and editing the Namokar Mantra book, to Mitchell Hall,

Martina Wohlfeld and Jan Markey for advising her, Jaya Carolyn Apolito for transcribing the tapes, to Saraswati Kathy McAdow and Maureen Lyons for typing the manuscript, to Pankaj D. Jain and Monica Jain for printing, and to all people who gave help and suggestions.

We are children of the Arihantas. May all beings be blessed with success, peace, health, self-realization and total bliss.

PART I
NAMOKAR MANTRA SCIENCE

INTRODUCTION

TRANSLATION

ṆAMO ARIHAṄTĀṆAṀ *I bow to the Arihantas, the perfected human beings, Godmen.*

ṆAMO SIDDHĀṆAṀ *I bow to the Siddhas, liberated bodiless souls, God.*

ṆAMO ĀIRIYĀṆAṀ *I bow to the Acharyas, the masters and heads of congregations.*

ṆAMO UVAJJHĀYĀṆAṀ *I bow to the Upadhyayas, the spiritual teachers.*

ṆAMO LOE SAVVA SĀHUṆAṀ *I bow to the spiritual practitioners in the universe, Sadhus.*

ESO PAÑCA ṆAMOKĀRO *This five-fold obeisance mantra,*

SAVVA PĀVAPAṆĀSAṆO *Destroys all sins and obstacles,*

MAṄGALĀṆAṀCA SAVVESIṀ *And of all auspicious repetitions,*

PAḌHAṀAṀ HAVAI MANGALAṀ *Is the first and foremost.*

See Pronunciation Guide, page 103.

18

THE NAMOKAR MANTRA is the essence of the gospel of Tirthankaras. These are the great Arihantas, the enlightened human beings who appear in decided eras, according to Jain chronology, to guide and teach mankind. An enlightened person has discovered the true nature of reality beyond appearances, and has penetrated all the secrets and mysteries of existence. This is the state of the Arihant which any aspirant may emulate. The Tirthankar, moreover, has created a spiritual path or bridge, a *tirth*, based on the principle of non-violence for the salvation of all souls. On this earth there have been twenty-four Tirthankaras, each of whom has reanimated and revolutionized the ahimsa path for each age. The most recent Tirthankar was Lord Mahavir, who lived 2500 years ago. His predecessor was Lord Parswanath.

In its eternal significance and application, the Namokar Mantra is without beginning or end. In their omniscience, the Tirthankaras have expounded upon the various states of consciousness as represented in the Namokar Mantra. They have taught the complete knowledge of the esoteric science of letters (Matrika Vidya) describing the special powers of each vowel and consonant. Subsequently, the Tirthankaras' chief disciples, known as *Ganadharas,* have used this precious knowledge to actually form a mantra by combining letters and their sounds and powers for the most beneficial effect.

Lord Mahavir and his disciples spoke Ardha-Magadhi Prakrit, the language of the common people of their region of what is now called the state of Bihar in India. Thus, he was able to touch their hearts in their own lan-

guage. By contrast, at that time aristocrats and literary
scholars conversed in Sanskrit, thereby excluding the
majority of the population. As the true purpose of language is communication, not the establishment of one class
as superior to others, the vernacular was the appropriate
form of expression for a universal spiritual message.

Understandably then, the Namokar Mantra has been
conveyed to this age in Prakrit, the language of its composition. For every era the Namokar Mantra has taken a
different form, but its essence has remained the same. As
only the *Ganadhar* is qualified to combine sounds as
taught by the Tirthankar, no one else can change the
form of the mantra without distorting it. There is a
deep, secret science to the combination of sounds.
Specific syllables are seeds for the awakening of latent
powers. Only the person who has been initiated into the
vibrational realms, who has actually experienced this
level of reality, can fully understand the Science of
Letters. Thus, the Namokar Mantra is a treasured gift to
humanity of inestimable worth for the purification,
upliftment and spiritual evolution of everyone.

Through a careful reading of this book you will be
empowered to understand the meaning and secrets of
mantra power. Etymologically, the word 'man' means
"mind" and 'tra' means "protection". This implies that a
mantra can fulfil any wish of the mind, that it can create
a happy and healthy mind. On a deeper level, a mantra
is a composition of divine sounds. The word 'mantra'
means "call", "invitation", "discussion". Through a
mantra we can contact the divine and awaken our own
sleeping powers.

Just as sound is the source of all manifestations in the
universe, a perfect combination of sounds can manifest

perfection in the spiritual aspirant. Mantra is used for realization of the true self, and the perfection of the soul. Ultimately, the true Self and mantra are one. The Namokar Mantra is actually the divine body of the Tirthankar.

When the Tirthankar is ready to discard the physical body and attain Nirvana and Siddhahood, in his love and mercy, he leaves behind for the universe his true eternal true body...his mantra body. When the power of the Namokar Mantra is completely awakened within us, then we can merge with the Lord.

CONCEPT OF THE ARIHANT

The Namokar Mantra appears with each Tirthankar. Without this Tirthankar (Arihant), the mantra is not available to us. This Arihant is God because he shows us the reality of the Siddha – the highest state of consciousness. In the mantra, the Arihant is the symbol of human perfection. He is the Godman.

In our human incarnation, we can achieve the state of Arihanthood. There is, however, a limitation to this state because of man's physical and causal bodies. The latter carries our accumulated karmic matter to the present existence. We cannot reach the unlimited state of Siddhahood until we have broken every bond with the physical body. Every man and woman has the potential to reach this state since the soul is perfect. Truth has no concern for caste or creed. At this time, twenty Arihantas exist. These divine souls can be contacted by anyone, at any time, by meditation and devotion.

The Arihant has laid the foundation for the spiritual system; he has guided our progress. For this reason, the Namokar Mantra first pays respect to the Arihant and then to the Siddha – the highest state of consciousness, our final goal of total liberation.

The Attainments

GENERAL BENEFITS

THE NAMOKAR MANTRA is the great protector and healer. The destruction of karma and attainment of knowledge and bliss are its spiritual benefits. Through the Namokar Mantra we can also gain various worldly attainments and *siddhis* (extraordinary physical and mental powers). Beginning the Namokar Mantra with *Om* will give worldly benefits. Through *Om* we contact cosmic energy whereby we can control the five elements and also affect our mental psychic abilities.

All form manifested in the universe has its root in the subtlest seeds of the basic elements – earth, water, fire, air and space. Chanting sounds such as *Om* or *Hreem*, will connect us to the ocean of universal energy, and, more specifically, to the subtle elements. This connection, when used with proper techniques, will actually attract worldly gain. It is basically a matter of calling on the subtle powers of nature to "create" the desired objective – a matter of the subtle elements manifesting form for us. But there are secrets involved in awakening specific powers or attainments: Only the guru can uncover these secrets for the disciple.

PURIFICATION OF KARMA

One may ask, how it is possible for a mantra to remove sins, as is claimed in the Namokar Mantra. To answer, one needs to understand that sin is simply negative karmic particles stored in the unconscious causal body. The mantra creates positive particles – divine particles. Through mantra power, negativities are destroyed and then the positive effects appear. First, we must understand the real nature of mantra. When sound and mind meet each other and merge, an electric current is produced. This current gives light. The first function of light is to destroy darkness. Then it radiates its brightness. It is the same with using mantra – the divine sound creates positive particles which first destroy the darkness of sin, and then grant gifts of 'light', worldly attainments and spiritual achievements.

The Namokar Mantra includes all *beej* mantras or seed sounds, such as *Om, Hreem, Arhum, Kleem, Shreem, etc.* Using specific seed sounds, we can contact the higher self, while others awaken powers and open chakras. Some are seeds of divine nectar, while still others work specifically to destroy negativities. The Namokar Mantra has the power to remove all poisons of sin in one instant and destroy the bad karma collected over hundreds of thousands of years in the causal body.

THE PATH TO LIBERATION

The main purpose of the Namokar Mantra is the achievement of *moksha* – a total liberation and freedom from the cycle of rebirth. This is the state of the Siddha. From the total enlightenment of the Arihant, the soul progresses to Siddhahood – complete liberation and boundless omniscience. By repeating the Namokar Mantra, all the divine qualities of the Arihant, Siddha, Acharya, Upadhyaya and Sadhu are awakened in the self, as well as the 'Three Jewels' – Right Knowledge, Faith and Conduct.

On the path to liberation, Right Knowledge, Faith and Conduct must be awakened. The Namokar Mantra assists in this process since it directly affects the 'Three

Jewels': *Namo Uvajjhayanam* is related to Right Knowledge; *Namo Siddhanam* is related to Right Faith; and *Namo Arihantanam, Namo Airiyanam* and *Namo Loe Savva Sahunam* are related to Right Conduct. The Sadhu is the practitioner striving for mastery of the inner self. By right practice – that is, Right Conduct – he can begin his journey in attaining this mastery, and the result is the next state – that of the Acharya. The Acharya represents perfect control, self-discipline and guidance for others in self-discipline. Control leads to self-mastery. The Sadhu who attains the highest result of Right Conduct through practice reaches the state of Arihant. In this case, Right Knowledge and Faith are also naturally attained.

The Upadhyaya is one who attains divine knowledge through direct experience. By this way he also attains Arihanthood. He is the symbol of knowledge – the Sadhu who has attained Right Knowledge and Right Conduct.

In the Namokar Mantra we pay homage to the five divine personalities, but they are not separate from us. They are actually symbolic of noble qualities, or states of consciousness, which we are striving to attain. They do not represent different paths to the goal of liberation but rather, various stages in the evolution of the soul. If we are spiritual practitioners, then in essence we are Sadhus, and we can progress to the ultimate states of Arihant and Siddha, and attain liberation.

REMOVING MENTAL CONFLICT

The Namokar Mantra is a great positive affirmation bringing us to a state of oneness with God...with the Arihant. We constantly collect positive and negative subtle material from thoughts, food, the company we keep and the environment. Simply by living in the world we collect these vibrations, many of which are detrimental. 'Dirt' will not wait for our invitation. There is much that we collect without any effort on our part. It is like the dust and grime that collect in our homes. However, we must make an effort to cleanse ourselves.

The mind can easily become filled with confusion, depression and sadness if it is not disciplined. The Namokar Mantra erases mental conflicts by creating permanent attitudes about our divinity and breaks identification with the lower animal nature.

When repeating the Namokar Mantra, which is basically an auto-suggestion, the mind consciously and unconsciously accepts the Arihant as its goal. When auto-suggesting, we are affirming only the positive. Do not think, "I am not bad." (Using the word 'not' will lessen the effect of our positive affirmation.) Rather think, "I am good. The Arihant is my goal, my true state. His qualities are mine." If our faith is perfect, and we repeat the mantra over and over again, while affirming our goodness, then we can reach the highest state of consciousness.

The various states of mind also relate to the chakras and Right Knowledge, Faith and Conduct. The conscious mind is the source of knowledge. The subconscious mind is the source of perception and faith. The unconscious mind is the source of conduct and energy. Consciousness corresponds to the centers above the throat. Subconsciousness corresponds to the centers below the throat to the navel. Unconsciousness corresponds to the centers from the navel to the base of the spine. Karmic particles are stored at the base of the spine. preventing the awakening of the kundalini. When the powers of the unconscious are awakened then the karmic obstruction is removed and we can realize our perfection. This brings us to the state of supercon-sciousness or *samadhi*.

Right Knowledge

Consciousness (Source of gross senses)

CHAKRAS: Throat center
Third eye center
Fontanel
Transpersonal point

Right Faith

Subconsciousness (Source of psychic powers; subtle senses)

CHAKRAS: Heart center
Solar plexus

Right Conduct

Unconsciousness (Source of conduct, energy and self-realization)

CHAKRAS: Navel center
Pelvic center
Root center

If we have attained Right Knowledge and Right Faith, then our power to discriminate right from wrong is strong, and it naturally follows that our conduct will be good. By repeating the Namokar Mantra we can purify the mind, awaken its powers and choose the right path.

PERSONALITY DEVELOPMENT

The personality is a reflection of thoughts, beliefs and actions. Through much repetition, the mantra becomes 'alive'. Its powers awaken within the practitioner who then experiences oneness with the Arihant and, subsequently, is filled with white light. The personality becomes very attractive and powerful. The divine sounds and pure colors fill the aura. One's aura will naturally affect anyone coming in contact with it. Just as we can feel repulsion, fear or tension when coming in contact with one individual's aura, so can we feel attraction, happiness and purity in another's aura. The Namokar Mantra purifies the aura colors. The animal mind can no longer misguide the practitioner. Rather, the Supreme Mind is guiding. Thoughts and ideals become very high, and the personality becomes greatly refined.

The Namokar Mantra is the key to control the lower animal nature and the development of the human and divine natures. Animal nature here refers to anger, greed, jealousy and fear. According to the Jain view of ontogeny, the soul evolves from mineral to plant to animal to human. We change from birth to birth, life to life. At present we have evolved to human states, but we are still attached to the animal nature. Without spiritual practice this animal nature cannot be overcome; this is represented by *Namo Loe Savva Sahunam*. *Namo Uvajjhayanam* stands for the Right Knowledge and understanding which will develop. *Namo Airiyanam* represents the control we will have over our nature. *Namo Arihantam* represents our enlightenment. The visualization of the white light of the Arihant removes the negative qualities of the animal nature. The 60th to 63rd verses of the *Rishi*

Mandal Stotram, which is a lengthy prayer in traditional verse form, explain how the complete visualization of the Jinabimbam will break the cycle of life and death for the practitioner. He or she will attain all virtues, perfection, and salvation.

In order to awaken our divine nature, more is required than the power of positive thinking. Thoughts take form as language and sound. The language of thoughts may be positive, but the corresponding letters and sounds may not be as useful. On the other hand, the Namokar Mantra is always helpful. It is a composition of divine sounds, a perfect arrangement of letters, the ultimate positive affirmation passed down to us from the highest souls.

Only with the help of mantra can one think positively all the time. Positive thinking can only take root when all negativity has been removed. The Namokar Mantra removes negativity and gives greater power of concentration, which helps in positive affirmations. We must visualize and merge with the Jinabimbam – the image of the Arihant – at the forehead. This will purify us. Then positive thoughts will come like the flow of a river.

AN EQUIVALENT TO FASTING

The ancient scriptures teach that repeating one mala (a rosary of 108 beads) of the Namokar Mantra will give the same effect as a one-day fast. In Sanskrit, *upavas* means to give rest to the stomach, to fast, to live with yourself and realize yourself. If we compare the effects of a fast to those of mantra repetition, we can see the similarities. The Namokar Mantra produces heat and energy which burn negativities and bad thoughts. This leaves a feeling of purity, lightness and peace. The mind attains a greater level of concentration, bringing us nearer to our true nature. The mantra actually brings benefits to us mentally, physically and spiritually.

The Great Healer

PROTECTING OURSELVES

THE NAMOKAR MANTRA represents the five colors, as do *Om, Hreem* and *Arhum*...white, red, yellow or orange, green or blue, and black.

White is the 'mother' or a blending of all colors. It shows purity, selflessness, cosmic consciousness and protective power. White is the color of *Namo Arihantanam*. The visualization of the Arihant and white light at the anterior fontanel will purify and protect.

Red is the great energizer. It activates energy and bestows total enlightenment. Red is the color of *Namo Siddhanam*, the color of the perfect divine soul. Visualize red at the forehead.

In ancient color science, yellow and orange rays were considered as one ray. Yellow shows wisdom, discipline and the power to choose a high goal in life. Orange is also a wisdom ray; it shows strong willpower. Orange and yellow are the colors of *Namo Airiyanam*, the Acharyas, who are the heads of congregations. They represent organized power, control and discipline. Visualize yellow or orange light at the right ear and the right side of the head.

Green and blue are also considered one ray. Both are the color of prana. Green is the color of balance and harmony. Blue, the color of truth and the power of speech, helps to develop spiritual and psychic powers. Thought energy can be transmitted by blue rays. Green and blue are the colors of *Namo Uvajjhayanam*, the Upadhyayas, who are the teachers of divine knowledge. They show us how to awaken powers and maintain balance of body, mind and soul. Visualize green or blue at the throat.

Black is the absence of color. It is receptive and consumes negativities. Black is the color of *Namo Loe Savva Sahunam*, the Sadhus, who are spiritual practitioners. The practitioner must destroy negativities before he can achieve success. Visualize black at the left ear and left side of the head.

Colors are also applied to the rest of the body. The right hand and arm as well as the right side of the torso are white. The left hand and arm as well as the left side of the torso are red. The left foot and leg are yellow or orange, and the right foot and leg are blue or green.

This entire color visualization is very powerful for protection. It should be done five times every morning. Use your right hand and touch the body parts as you apply the mantra with the appropriate color visualization. You can bring more energy to the navel if you adjust the system slightly. After you apply the colors to your head and neck, use your left hand and with a sweeping motion apply white from the fingertips of the right hand, running up the entire arm and down the right side of the torso to the navel. Then use your right hand and with the same sweeping motion, apply red from the fingertips of your left hand up your arm and down the left side to the navel. For the legs, use your right hand and bring energy up with a sweeping motion from toes to navel – first the left side in yellow or orange, then the right side in blue or green.

Another system of applying the Namokar Mantra to the body:

Lord Parshvanath

Lord Adinath

Top of the head (white)

Om Namo Arihantanam
Shirsham Raksha Raksha Swahah
Protection for the head

Face (red)

Om Namo Siddhanam
Vadanam Raksha Raksha Swahah
Protection for the face

Heart (yellow)

Om Namo Airiyanam
Hreem Hridiyam Raksha Raksha Swahah
Seed of all extraordinary powers;
protection for the heart

Navel (blue)

Om Namo Uvajjhayanam
Hroum Nabhim Raksha Raksha Swahah
Seed to purify sushumna and contact divine;
protection for the navel

Feet (black)

Om Namo Loe savva Sahunam
Hrah Padou Raksha Raksha Swahah
Seed of power to destroy obstacles;
protection for the feet

THE HEALING TECHNIQUE

Everyone is endowed with healing power to help himself and others. Sometimes though, this energy is lacking due to the effects of certain foods, negative karma, accidents, and imbalances in the system. There are many reasons for the loss or lack of energy and for illness, but a simple definition of disease is a blockage of energy.

Negative thoughts play an important part in creating blockages. Fifteen minutes spent in anger produce toxins which will last six months in the body. The root of many diseases is mental. The Namokar Mantra has great power to clean the mind and dissolve blockages. According to the Matrika Vidya, the *'Nam'* in the word *'Namo'* is a painkiller. If you repeat the Namokar Mantra in times of pain, that pain will be greatly lessened or stopped. Also, the Namokar Mantra bestows its five colors, relating to the five pranas, each with particular healing powers for various illnesses.

Prana, the vehicle of healing, is the vital current of life force manifesting in all phenomena on the physical and mental planes. It is the source of all energy. Prana is freely available in the air we breathe. According to medical science, when inhaling, we take in 20.96% oxygen, 79% nitrogen and .04% carbon dioxide. However, according to the yogic system of human development, we are further concerned with taking in pranic energy. While prana is one basic energy, it takes five forms – prana, apana, samana, udana and vyana. These five subtle airs govern all functions of the body. If they are not working properly, disease will develop.

Prana is blue or green in color; it controls respiration. Apana is black and controls excretion. Gas problems in

Method of Protecting the Body by the Namokar Mantra

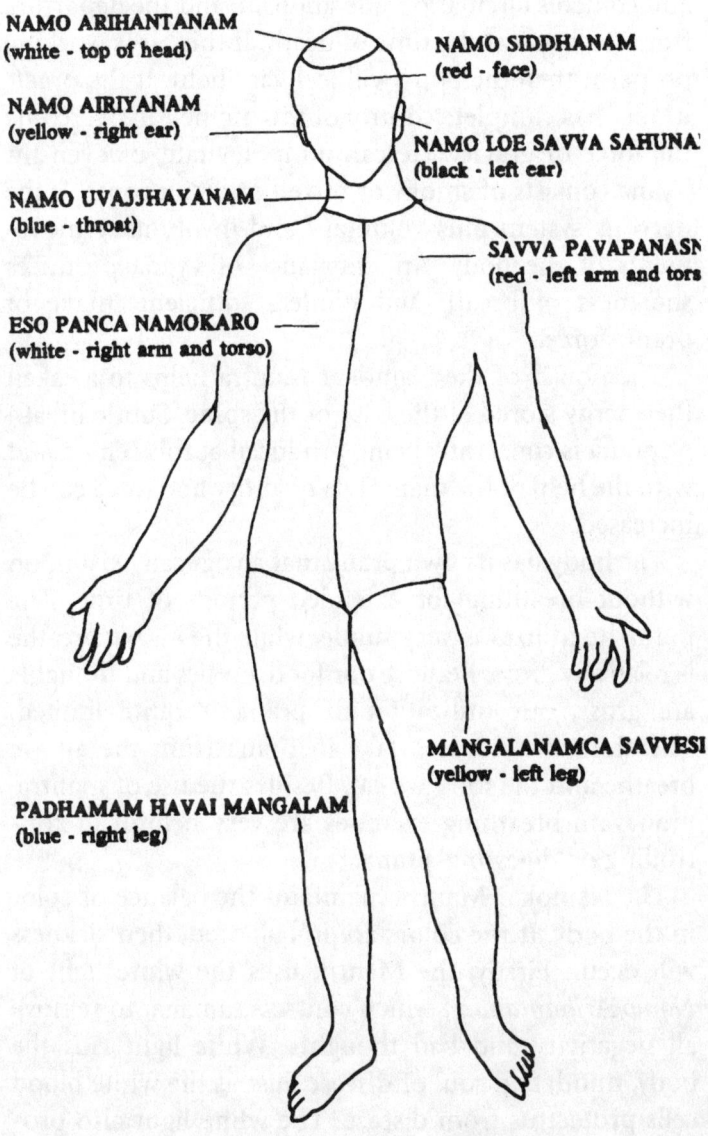

NAMO ARIHANTANAM
(white - top of head)

NAMO SIDDHANAM
(red - face)

NAMO AIRIYANAM
(yellow - right ear)

NAMO LOE SAVVA SAHUNA'
(black - left ear)

NAMO UVAJJHAYANAM
(blue - throat)

SAVVA PAVAPANASN
(red - left arm and tors

ESO PANCA NAMOKARO
(white - right arm and torso)

MANGALANAMCA SAVVESI
(yellow - left leg)

PADHAMAM HAVAI MANGALAM
(blue - right leg)

the body are caused by imbalanced apana. Samana is white; it controls digestion and assimilation. Udana is red and controls circulation, metabolism, and the departure from the body at the time of death. If this air is working properly, then the body will feel very light. If the practitioner has complete control of this air, he can overcome the force of gravity. He can jump, levitate, or even fly. Vyana consists of smoky or mixed colors; it controls the nervous system plus voluntary and involuntary movements of the body. An imbalance of vyana produces shortness of breath and hinders sufficient intake of pranic force.

The power of the Namokar Mantra helps to awaken the energy stored at the base of the spine. Subtle breath or prana is constantly being produced at this center, and with the help of the mantra its quantity and force can be increased.

The body has its own prana that a yogi can exist upon without breathing for extended periods of time. The prana he utilizes is very subtle, while the air we breathe is relatively gross. Because our food, bodies and thoughts are gross, our utilization of prana is quite limited. Therefore, we need to take in prana from the air we breathe and the food we eat. Besides the use of mantra, pranayam breathing exercises are very helpful in controlling or 'digesting' prana.

The Namokar Mantra maintains the balance of color in the body. If the colors are imbalanced, then sickness will occur. Firstly, the Mantra uses the white light of *Namo Arihantanam,* which controls samana, to remove all negativity and bad thoughts. White light rids the body, mind, and soul of disease, just as the white blood cells protect us from disease. The white light also pro-

tects us from psychic attacks. It is necessary though, to
visualize the figure of the Arihant in white light at the
fontanel. —

Namo Siddhanam, which controls udana, is red. Its
place is the third eye center. Red light governs the vitality
of the body, particularly the creative, procreative and
restorative processes. It causes red blood cells to multiply
and invigorates circulation. The purest red light is the
color of the Supreme. The lower red shades – scarlet and
red mixed with brown – are colors of pride, anger and
desire. In order to control these desires, the lower shades
must be controlled. The pure red color also removes our
unconscious karmic problems.

Namo Airiyanam, which controls vyana, is yellow or
orange. Yellow controls digestion and elimination by the
intestines and liver. It strengthens the nervous system
and awakens the reasoning faculties. Yellow maintains
the health of the solar plexus and the brain. If the brain
is healthy, all body functions will work properly. Yellow
also maintains the health of the spine. if the spine and
joints are lubricated by sufficient 'yellow fluid', back
problems and arthritis will be prevented. Orange assists
in circulation and assimilation.

Namo Uvajjhayanam, which controls prana, is blue or
green. Green nourishes the heart chakra. It has soothing
effect on the nervous system and promotes general harmony of the body and mind. Blue nourishes the throat
chakra. It is cooling, soothing and astringent. It can cure
fevers and stop bleeding and various pains. Blue and
green have great healing powers since they possess the
color of prana.

Namo Loe Savva Sahunam, which controls apana, is
black. Black absorbs negativities and provides a 'black

hole' into which the practitioner can focus, merge and dissolve the self and reach the transcendental state. This 'black hole' can be created by visualizing a mass of small black dots in space into which all negativities, from the big toe to the top of the head, can be absorbed.

We use the Namokar Mantra as the basic mantra for healing. Not only do we apply the mantra on the body of the patient, but we must convert our own bodies into the mantra. (We gradually attain this ability by repeating the mantra 108 times daily.) When using the mantra to heal, you must first feel that it is a part of your life force. Do not identify yourself egotistically as the healer. Rather, feel that it is the power of the mantra that heals. You are only a medium. When you apply the mantra in this state of mind and surrender completely, the healing energy will be effective.

Apply the mantra three times visualizing colors on the patient's body. Feel the energy and vibration while focusing the energy from your right hand to the affected area of the patient's body. Using the specific part of the Namokar Mantra and its related color, localize the healing energy and apply it to the affected part. Visualize and feel that the problem is going away. Contact the cell groups and 'cell-mind' of the affected part and send the suggestion, "You are Arihant. You are Siddha. You are Divine. Remove all imbalances and work properly." The disease will be removed.

This is the secret power of the Namokar Mantra; it can be used anytime to heal yourself or others. The power and knowledge for healing is available anytime. Still, we need a guru to teach us the special techniques and to awaken the power. The guru discloses all secrets.

No healing system should be started without the guidance of a guru. Why do you go to school? Why do you send your children to school? You can just as well buy the books and study on your own and become a master. Correct? No. It is necessary to contact a living guru. Everyone is a healer but you should only awaken this power with proper guidance.

Now you understand that by color, pranic force, sound and inner body systems, the Namokar Mantra has great healing powers.

COLOR SCIENCE

The Purest Ray

When we want the purest result from the contemplation of color, then we must contact the Tirthankaras. Their true essence is color. If the source of the color is pure then the color itself will be pure. From the Tirthankaras come the most highly refined vibrations of color by which one can contact the highest Self. The attainments of the Tirthankaras are reflected in their colors. These colors are available to us at all times. Merging with a specific color and vibration helps us awaken all the divine powers of that color within us.

To awaken the highest attributes and characteristics of a particular color, we must first contact a Tirthankar with that corresponding color, adopting the attitude of *Namo* meaning *Namaskar* or "I bow". After preparing ourselves with humility, we can meditate and concentrate on the divine form of the Tirthankar in his lotus posture and on his particular color.

Generally, if we concentrate on the color black, this can produce a negative result. But concentrating on the

Tirthankaras who possess black color will give us the highest result of black. We can thus remove and destroy our negativities. The single color attributed to each Tirthankar came about for various reasons including the subtle or pronounced color of his skin and the appearance of the color in this aura. The skin's pigment, incidentally, is tinted by the presence of minerals and subtle metals, such as gold, copper, iron, *etc*, which are naturally present in the body.

We can also visualize the symbol of each Tirthankar, for more specific contact, for example, the cobra for Lord Parswanath and the lion for Lord Mahavir.

Color-Breathing

Color-breathing is very useful for self-healing, for energizing and balancing the colors in the aura and subtle body.

The following technique is especially beneficial when performed outdoors facing the sun:

Raise your arms and hands straight up with fingers outstretched. Inhale deeply and imagine that the breath is filling the body with the specific color that is suitable for you, while energy is streaming in through the fingertips. While holding in the breath for a few seconds, keep your arms straight but lower them in front of you, parallel to the ground. Then exhale while lowering the arms to your side. Repeat this process about ten times.

You can perform another simple technique *anywhere* while either sitting erect or lying down in the 'corpse' pose.* Inhale deeply and imagine that with each breath

---

* Corpse pose – Savasan: *This is the pose of total relaxation performed while lying on the back with spine aligned. Arms are straight yet relaxed, not touching the body, with palms facing upward. Legs are straight yet relaxed, not touching each other.*

you are bringing in your chosen color. Either you can fill your entire body with this particular color or direct the breath with color to a particular chakra. Hold the breath for a few seconds; then exhale. Imagine that all negativities and obstacles are leaving that chakra (or the entire body), exiting with a smoky color. When you direct the color to a chakra, the color will automatically give help wherever needed in the body.

Another color variation is the rainbow technique. While inhaling, imagine that the spectrum of all colors is entering the body like a rainbow. This will help to balance all colors simultaneously.

The Chakras and their Colors

There is a complex relation between color and the chakra system. The colors appear due to their powers, frequency of vibration, letters (sound), elements and pranas. There is not only one color for each chakra. However, here we will suggest a simple, useful system for color-breathing:

> Muladhar is nourished by yellow.
> Swadhisthan is nourished by white.
> Manipur is nourished by red.
> Solar plexus is nourished by yellow.
> Anahat is nourished by green.
> Vishudha is nourished by blue.
> Ajna is nourished by red or blue.
> Sahasrar is nourished by white and all colors.*

---

* See page 57 for diagram of location of chakras.

Treatments by Color for Illnesses

Within the Namokar Mantra healing system, after applying the mantra to the body, you may use the following colors for treatment of specific problems. Also, for self-healing, you may use them with color-breathing.

acne blue, indigo
anemia red
asthma orange, red
nosebleed indigo
bruises blue
burns blue, turquoise
cough (wet) orange; (dry) indigo
chicken pox blue
colds (head cold) green; (loosens colds) yellow
constipation yellow, orange
cramps orange
diarrhea violet, blue
ear infection blue
emotional disturbances pink, violet, white
energy builders red, orange
fever blue, green
hay fever green, blue, violet
headache green, blue, turquoise
hiccups orange
insomnia blue, indigo, violet, green
measles blue
confusion yellow
nausea and vomiting blue
nervousness orange
general pain blue, white
sore throat blue, indigo
sunburn blue, turquoise
toothache blue

The Science of Letters

THE MATRIKA VIDYA

THE MATRIKA VIDYA is the Science of Letters, describing the special powers of each vowel and consonant. There are 16 vowels and 33 consonants in Sanskrit. Representatives of all matrikas (or letters) are included in the Namokar Mantra. The consonants and vowels combine to produce *seed* sounds which have miraculous powers.

The term 'matrika' is derived from the word 'matra' meaning "measurement" – in this case, measurement of time. For example, it takes one complete *matra* to utter any vowel, and half a *matra* to utter any consonant. The measurement grows with 'stretched' sounds. The length or measurement of uttered sound is categorized into three levels with three different results. Repeating a short sound (as in *hrum*) will effectively remove karmic particles. Repeating it doubly long (as in the extended vowel sound in *hroom*) will facilitate worldly attainments and spiritual achievements. Repeating an even more extended sound will change the vibration or atmosphere of the environment.

When we understand the Science of Letters, we can see that a mantra is formed with so many considerations in mind. The *mantra* is only one aspect of the power of the letters.

The Matrika Vidya is the base of all worldly and spiritual knowledge. Each letter holds a wealth of knowledge, which it has collected since the beginning of time. If we know how to contact the power of one sound, then all its history will be opened to us. It would be as if we had opened a cosmic encyclopedia for a particular letter, and all of its mystical knowledge would appear before us. Its secrets and mysteries would be realized. When this knowledge is awakened, then we can realize true and ultimate reality.

If modern civilization should meet with a great catastrophe and destruction, then a mantra such as the Namokar Mantra could preserve the full science of sound, letters and language. All the secrets of the mantra would remain but would have to be realized through practice. Any knowledge which may be lost through such a fundamental planetary change, can be regained by the preservation of language through the mantra. Civilization is based on language. The science of the universe, the soul, the elements, color and health, the theories of creation, action, and destruction – all the advancements of man can be preserved.

Sound is infinite. Sanskrit and other languages have given form to sound. I am from India; my background is in Sanskrit, so naturally I am teaching this system. But I am not excluding any other forms of language or sound. Sound is sound. But we can easily understand the science of sound by the ancient Matrika Vidya, since the Realized Souls from the beginning of mankind stressed this knowledge as a complete system for self-realization.

The body, the universe and the powers of letters are closely interrelated. The main powers governing the microcosm of the body are sun, moon and fire. This is also true of the macrocosm as well as of letters or sounds. The following chart summarizes these correlations.

LETTERS	POWERS	STATES OF MIND	CHAKRAS
Vowels	Moon	Consciousness	Throat, Third eye, Fontanel, Transpersonal point
Consonants	Sun	Subconsciousness	Heart, Solar Plexus
Semi-Vowels	Fire	Unconsciousness	Navel, Pelvic, Root

The Namokar Mantra is composed of vowels, consonants and semi-vowels in perfectly balanced order, imparting balanced vibrations from sun, moon and fire energies – affecting the body through the chakra system, the mind through the three levels of consciousness, and the soul.

BASIC COSMOLOGY

According to the *Jain Yogashastras* of Acharya Hemchandra (11th century), the universe is a manifestation of energy, sound and shape. On a subtle level, sound at ultra-high frequencies is the combination of color and light. Without sound there is no light. Likewise, without light there is no sound. Wherever there is light there is subtle sound which can be heard by the adept spiritual practitioner. Similarly, this subtle sound can be seen as light. Each letter represents a particular sound. And sound is a combination of color and light. Color is the female force in nature; light is the male force. Color is light vibrating at various frequencies.

The *Yogashastras* elaborate that color represents the five elements – earth, water, fire, air and space – each of which having its corresponding shape or *mandal*. In their subtlest form, these shapes are configurations produced by a magnetic field relating to the specific frequencies of sound and light. The different frequencies of energy manifest as the various shapes of the elements.

ELEMENT	COLOR	SHAPE
Earth	Yellow	Square
Water	White	Crescent
Fire	Red	Triangle
Air	Blue	Hexagram
Space	Smoky or Mixed	Circle

Shape represents energy, and sound represents action. Without sound there could be no creation, preservation or destruction. So, the universe, which is made up of earth, water, fire, air and space, is the result of sound. In the Bible, John says, "In the beginning was the word and the word was with God, and the word was God... All things were made by Him [the word], and without Him was not anything made that was made."* If we knew the secret of sound, we could control the elements and all powers.

OM: THE ORIGIN OF MANTRAS

The twelfth verse of the *Shraman Sutra* describes how *Om* is the essence of the Namokar Mantra, '*Arihanta Asrira Airiye Uvajjhaya Munino Panchakhar Nippano Omkaro Panchparamitthi*'.

The initial letters of the names of the five divine personalities combine to make *Om* or *Aum* according to the following formulation:

Arihant A
Ashariri (Siddha) A
Ācharya Ā
Upadhyaya U
Muni (Sadhu) M
A + A + Ā = Ā: 3 A sounds combine to make Ā
A + U = O: A and U combine to make O
O + M = OM: or *Aum*

---

* *John, 1: 1–3*

Om is the base. It is the mother and source of all sounds. When chanting *Om*, we are also chanting the Namokar Mantra. When chanting the Namokar Mantra, *Om* is there, everywhere. *Om* is the original sound. *Om* represents the three main forces in nature: A stands for creation, U for preservation and M for destruction.

According to the particular power of the sound, or the combination of sounds, a mantra will have an effect of creation, preservation or destruction. The Namokar Mantra basically uses the destructive force since without the destruction of bad karma nothing beneficial can be attained. *Namo Arihantanam, Namo Airiyanam, Namo Loe Savva Sahunam* have destructive power. The Namokar Mantra also includes the powers of creation and preservation. Through the mantra, one can attain divine knowledge and samadhi, which is the blissful state of self-realization. The divine knowledge of *Namo Uvajjhayanam* is representative of creation. And the samadhi of *Namo Siddhanam* represents preservation.

Om can be used as a shorter form of the Namokar Mantra, as can *Arhum* or *A, Si, Ā, U, Saya Namah.*

THE ARHUM YOGA LOGO

In order to describe the meaning of the Arhum Yoga logo, we must describe the meaning of *Om, Hreem* and *Arhum.* These three interrelated sounds symbolize the techniques and the aim of Arhum Yoga. They symbolize knowledge of the external, the internal and the Supreme. They are Right Knowledge (*Arhum*), Right Faith (*Om*) and Right Conduct (*Hreem*).

Visualize the picture of *Hreem, Om, Arhum.* First on the outside is *Hreem;* this is the universe, nature and the five elements. Embedded in this lies *Om* whose positive energy moves in one direction to the Supreme, completely and wholeheartedly. In the center there is an unmoving flame of fire burning brightly. This is *Arhum*, the power of the soul. It is the Arihant, the soul in its perfected state.

The meaning is that you are the universe or *Hreem*. Your entire energy is *Om*, and your soul realizes its highest state of perfection, *Arhum*. By this system you can know the world, you can see how energy moves and you can see your Self, how all powers awaken. Watch and awake yourself.

Sound is the base of this knowledge. When you see Reality, then you will understand that this is the complete picture of the Supreme – omniscient, omnipresent and omnipotent.

Having spent most of my life experimenting with sound, I have found three sounds to be very powerful – *Om, Hreem* and *Arhum*. *Om, Hreem* and *Arhum* are the essence of the Namokar Mantra and manifestations of the Arihant; they should be repeated daily. The Namokar Mantra is an elaboration of *Om, Hreem* and *Arhum*.

Lord Chandraprabhu

Lord Mahavir

THE POWERS OF OM, HREEM AND ARHUM

Om is the seed of God. It is the seed of our entire energy moving upward to meet the Supreme. *Om* is the mother of all sounds. *Om* has all powers of the matrikas. By repeating *Om* we open and clean our arteries which will prevent circulatory problems. *Om* is used mainly for vibrating the top of the head – the transpersonal point. At this point, the energy enters into the body and with a spiralling movement passes through to the feet. The *Om* sound will especially affect the portion between the navel and the top of the head.

Many religions have accepted the importance of sound vibration and use it systematically to affect the consciousness of practitioners. The Muslims chant *Allah*. This vibrates the root of the navel. *Wah Guru*, which the Sikhs chant, also gives the same effect, as does *Jehovah*. Buddhists use *Om Mani Padme Hum* to awaken the energy centers. Generally, the Hindus (and other sects) use *Om*. Muslims use *Amin* and Christians use *Amen*, both of which are forms of *Om*.

Om is the universal sound. We can hear it in the first cry of a baby, in the sound of the wind and the ocean, in the roar of the lion. *Om* has three powers – *A, U* and *M* (another way of spelling *Om* according to Sanskrit). As previously mentioned, *A* represents creative power, *U* represents preservation and *M* represents destructive power.

Om shows us the world as well as our destiny. It takes us from the three *Gunas* – inertia, activity and purity – to the highest state beyond thought, beyond sound. *Om* represents this highest state. We call it God, Absolute, Arihant, Siddha. We can call it the highest state of consciousness. It brings us to total transformation – complete merging with the divine. *Om* is the original sound. It really is not derived from any language, but is antecedent to all language.

Arhum is the seed of the soul in its perfected state. the Rishi Mandal Stotram explains in the third sloka, *Arhum Ityaksaram Brahma Vacakam Parmasthinah*..., that *Arhum* is the "All." Is is the symbol of the five main powers – Arihant, Siddha, Acharya, Upadhyaya and Sadhu. *Arhum* is the seed of the Siddha chakra, the 'wheel' of the Siddhas, emanating the five divine rays of color. *Arhum* purifies the mind and carries the practitioner from a gross level to the higher self. *Arhum* is also considered the seed sound of the Goddess Saraswati. All the powers of the matrikas are included in *Arhum*. It is the seed of knowledge and sound manifested as speech.

The main purpose of *Arhum* is to vibrate the sushumna, which is the main nerve or nadi in the center of the spine. *Arhum* will open and clear the sushumna and prepare the passage for the kundalini (concentrated life force) to ascend. *A* is the seed of the moon, *H* is the seed of the sun and *R* is the seed of the power of fire. The letters affect the ida, pingala and sushumna nadis. The nadis are subtle nerve tubes which carry pranic energy. The ida which is ruled by the white moon energy, spirals up the spine from the left gonad (ovary or testis) to the third eye center. The pingala is ruled by the golden sun energy and spirals up the spine from the right gonad to the third eye center.

The nadis are purified and energized by the force of *Arhum*. *Arhum* has various forms. *Arhoom* vibrates the fontanel. *Hum* vibrates the throat. *Hrh* vibrates the solar plexus. *Rhah* awakens the navel. *Rhara* is a very powerful form of *Arhum*. This sound can completely awaken the sleeping kundalini. Its force hits the kundalini power and creates heat which pushes the energy in the sushumna. *Rhum* is used for inner purification with pranayam (breathing exercises) and visualization. these various sounds are all part of *Arhum*

Hreem is the seed of the universe. The universe is a combination of the elements and their corresponding colors and pranas, and *Hreem* controls these. *Hreem* represents the form of the universe and the form of the human body. By *Hreem* we can connect with the subtle forces and seeds of the universe. Because of this *Hreem* can bring material gain and the fulfillment of worldly needs. *Hreem* represents sun energy. Chanting it will create heat at the base of the spine. This heat can help to awaken the kundalini. By chanting *Hreem,* we will increase our strength and energy. *Hreem* also represents the twenty-four Tirthankaras by its five colors. Verses 22 to 26 of the Rishi Mandal Stotram describe this in detail.

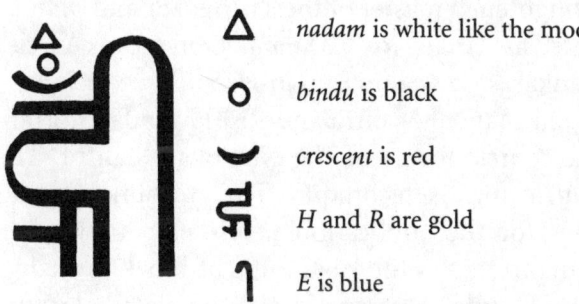

△	*nadam* is white like the moon
○	*bindu* is black
⌣	*crescent* is red
ह्रीं	*H* and *R* are gold
ी	*E* is blue

The stotram describes the colors of the Tirthankaras and where they sit in *Hreem:*

Chandraprabhu and Pushpadant are white in the *nadam*. Neminath and Muni Suvrat are black in the *bindu*. Padamprabhu and Vasupujya are red in the crescent. Parswanath and Mallinath are blue in the *E*. All others are gold in the *H* and *R*. This system shows how we receive divine color from its divine source.

THE RISHI MANDAL MANTRA AND STOTRAM
Their Relationship to the Namokar Mantra

The Namokar Mantra is the source of all seeds of mantras and combined seeds (*kootasth beej*). Based on the Matrika Vidya, it represents all groups of letters – vowels and consonants. The Namokar Mantra is the mother of all mantras. So in a subtle way, when we are repeating my mantra, the Namokar Mantra is there. All mantras of the belonging to the creative, preserving or destructive types are related to the Namokar Mantra. Also, the Namokar Mantra can destroy any negative effects of black mantras which are malevolently uses sound patterns intended to harm.

The Rishi Mandal Mantra is a commentary on the Namokar Mantra and praise of the divine self. (*Rishi* means enlightened master of the Tirthankar and *mandal* is translated as "circle". *Rishi Mandal,* then means "circle" of Tirthankaras, as symbolized in *Hreem*).*

The Rishi Mandal Mantra and Rishi Mandal Stotram are white *tantric* forms of the Namokar Mantra. The term *tantric* means technique. The stotram includes techniques for the application of esoteric knowledge, and the mantra gives the key points of this knowledge. The Rishi Mandal Stotram is a more detailed commentary describing the system of the twenty-four Tirthankaras and how they are related to letters and colors. It also includes complete knowledge about the conscious, subconscious and unconscious mind, and techniques for health and body and mind, worldly attainments and spiritual salvation.

The mantra and stotram stress *Om, Arhum* and *Hreem,* which also represent the Namokar Mantra. *Hreem* is widely employed in the tantric system, using color and sound. *Hreem* represents universal energy. It is derived from *Arihantanam – ri* and *han* combine to form *Hreem.* The five colors of *Hreem* represent the five stages of consciousness. These five stages are also represented by *Hram, Hreem, Hrum, Hroom, Hrem* in the Rishi Mandal Mantra. These five sounds, plus *Hreim, Hroum,* and *Hrah* for Right Knowledge, Faith and Conduct, are the eight forms of *Hreem.*

The eight seed sounds of *Hreem* in the Rishi Mandal Mantra awaken the seven chakras and related centers.

Hram	Sahasrar (top of head)
Hreem	Ajna (third eye center)
Hrum	Visudha (throat center)
Hroom	Anahat (heart center)
Hrem	Manipura (navel center)
Hreim	Swadhisthan (pubic center)
Hroum	Muladhar (root center)
Hrah	Ganesh (anal center)

The Rishi Mandal Mantra also includes the first letters of each of the five divine personalities:

A	Arihant	*U*	Upadhyaya
Si	Siddha	*Sa*	Sadhu
Ā	Acharya		

It then stresses the attainment of *Gyan, Darshan, Charitra* (Right Knowledge, Faith and Conduct). The Rishi Mandal Mantra begins with *Hreem* and ends with *Hreem,* finally stating that all powers come from *Hreem.*

See page 17 for Tirthankaras set in Hreem Symbol.

With the repetition of the mantra and stotram, Right Knowledge, Faith and Conduct awaken; these purify the physical, fire and causal bodies, respectively. The stotram gives the key for merging with the Arihant and for the mastery of the inner bodies by visualization of mantra and color.

Rishi Mandal Mantra

OṀ	HRAṀ	HREEṀ
transpersonal point	*fontanel*	*forehead*
white	white	red

HRUṀ	HROOṀ	HREṀ	HREIṀ
eyes	*nose*	*mouth*	*throat front*
yellow or orange	blue or green	black	white

HROUṀ	HRAṀ
heart to navel	*navel to feet*
yellow or orange	blue or green

A, SI, Ā, U, SĀ

SAṀYAG JNĀN DARSHAN CHĀRITREBHYO
HREEṀ NAMAH SWĀHĀH

Sound and the Serpent Power

PURIFYING THE CHAKRAS

THE CHAKRAS ARE CENTERS of whirling energies within the subtle body. They correspond with major nerve plexuses in the physical body. The chakras are directly related to the different levels of awareness in man from the instinctual to the spiritual. They control the flow of pranic energy like transformers, since they vibrate at different frequencies.

Sound, light, the five elements and the five pranas are all interrelated with the chakras. These influences show .us the various mantras, colors and shapes of the chakras. The lotus is the symbol of the chakras with varying numbers of petals, letters within those petals, colors, shapes within the center of the lotus, and a main seed sound or *beejakshara*.

By chanting these seed sounds, we vibrate and purify the chakras and the nadis and allow the kundalini to rise from the base of the spine to the top of the head. The sound vibration balances the influences interrelated with the chakras and removes blockages and impurities. It also benefits the physical body. For example, chanting *Hum* will not only affect the throat chakra, but also any physical problems relating to the throat itself. Any part of the subtle or gross body may be purified, healed and awakened by the use of sound vibration.

SEED SOUNDS AND THE CHAKRAS

OM Sahasrar (top of head, thousand petalled
lotus, transpersonal point)
ARHOOM center of head (fontanel, pineal gland)

EIM   Ajna (third eye center between eyebrows, pituitary gland)

HUM   Vishudha (throat center, thyroid gland)

YUM   Anahat (heart center – at level of sternum, in center of chest)

RUM or Manipur (navel center – root of the navel,

RAHAH three fingers distance below navel, set within body)

HRAH or HROUM Solar Plexus

VUM   Swadhisthan (pubic center – at level of pubic bone just above penis or clitoris, directly in front of coccyx)

LUM   Muladhar (root center – perineum, set within the body at the cervix in women, between the scrotum and anus in men)

GUM   Ganesh (anal center)

ARHUM Sushumna (central nerve in spine)

HREEM To awaken kundalini

CHOOSING THE CORRECT SOUND

Each chakra has its true "owner", such as *Yum* for the heart chakra. But various other sounds can still affect the chakra. *Hum*, for example, is the seed for the throat, but it also is the master key for all subtle sounds. *Hum* has the power to break the *brahma, vishnu* and *rudra granthis*. These are knots located at the muladhar, anahat and ajna chakras preventing free movement of the pranic or kundalini energy in the sushumna. *Hum* also destroys negativities. Arhum is a beautiful divine sound. *Hum,* which is a part of it, can be used anywhere. *Hum* will awaken any center where it is visualized. *H* represents the sun energy, pingala and sushumna. Hreem and Arhum and its variations will also affect the heart.

The following diagram shows general spinal and frontal body
locations of the chakras:

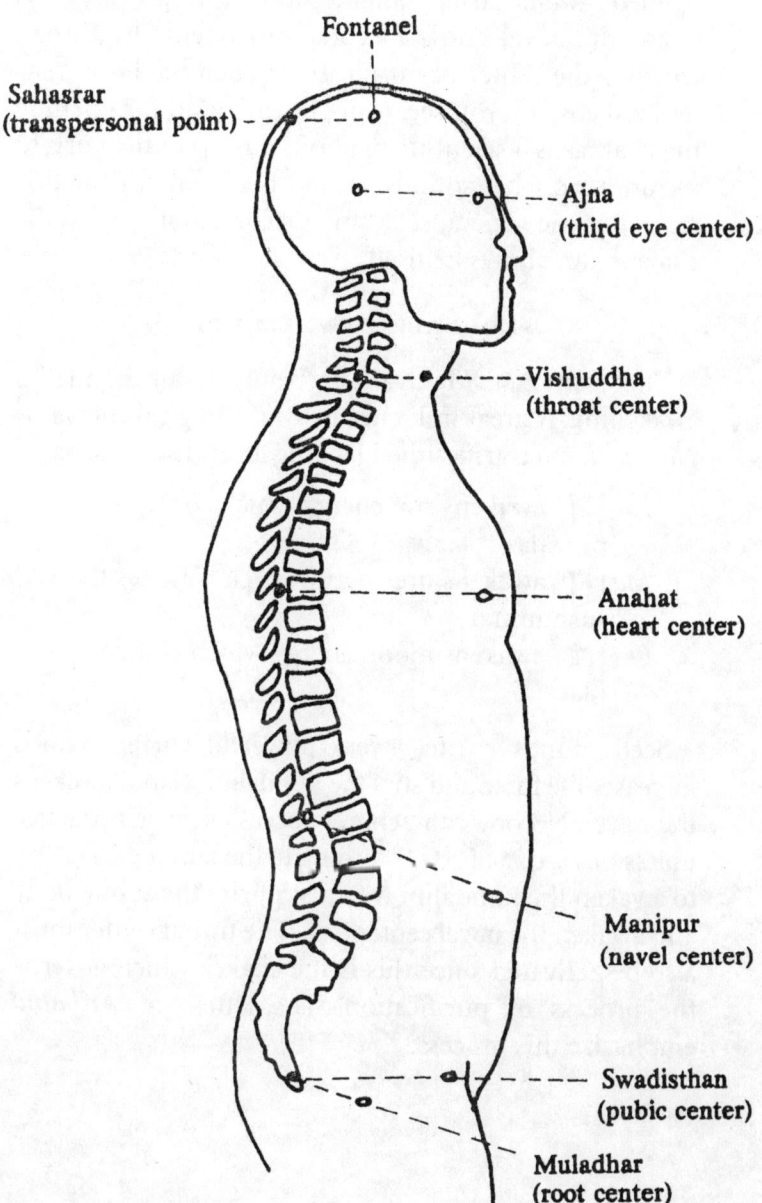

According to the Namokar Mantra system, the navel
chakra must first be awakened before the heart can be
opened. *Namo Arihantanam* shows this process – *ri*
opens the navel and *han* (or *hum*) opens the throat,
which is the center of purification. Then the heart center awakens. So, you see, using inner sound for opening
the chakras is a scientific method. It is up to the guru to
recommend what sounds are most suitable for the disciple. But the general seed sounds corresponding to the
chakras can always be used.

KUNDALINI AWAKENING

The Namokar Mantra can bring about kundalini
awakening. A great influence in awakening this power is
Namo. *Namo* purifies the ida, pingala and sushumna.

> *Na* [ण] awakens sun energy which controls the
> pingala.
> *Ma* [म] awakens fire energy which controls the
> sushumna.
> *O* [ओ] awakens moon energy which controls the
> ida.

Seed sounds create a very powerful current which
increases life force and stirs the kundalini. Namo awakens
the navel. No one can achieve success in yogic practice
unless this is done first. According to the Jain yogic system,
to awaken the kundalini we must purify the subtle body
and awaken the navel center first. The throat center must
also be activated since this is the chakra which governs
the process of purification. The letters of *Arihanta*
emphasize this process:

> *A* = opens the throat chakra
> *Ri* = opens the navel chakra
> *Han* = opens the throat chakra
> *Ta* = opens the navel chakra

Na creates a current which opens the palatal center. This is the door through which the kundalini can reach the fontanel. The fontanel is also the place of udana air. This air gives help to awaken the kundalini; it is always moving upward. When udana is controlled, one can levitate, control the subtle body and awaken great psychic powers.

The sound *O* also plays its part and creates an upward flow of energy. The sound *Nam* repeatedly found at word endings in the Namokar Mantra, opens the passage of the sushumna by removing blockages and negativities.

The Namokar Mantra, as I have said before, represents the three main powers – creation, preservation and destruction. The Namokar Mantra also represents the three *gunas* (aspects of nature) – purity, activity and inertia. Inertia predominates in the lower chakras, activity predominates in the middle chakras, and purity in the higher chakras. Therefore, the Namokar Mantra symbolizes the progressive rising of the kundalini through the chakras. But it is more than a symbol, it is an extraordinary combination of divine sounds, and if you repeat it accurately, your kundalini power will start to work.

The letters of the Namokar Mantra and their corresponding chakras:

NUMBER CODE: 1 throat center   2 heart center
3 navel center   4 pubic center   5 root center

3   4   1 3   1   3   3
NA MO   A RI HAN TA NAM

3   4   5   3   3
NA MO   SI DDHA NAM

3   4   1   2   3   2   3
NA MO   AI YA RI YA NAM

3   4   1   4   2   2   3
NA MO   U   VA JJHA YA NAM

3   4   5   1   5   4   5   1   3
NA MO   LO E   SA VVA SA HU NAM

1   5   3   2   3   4   2   3
E   SO   PAN CA   NA MO KA RO

5   4   3   4   3   3   5   3
SA VVA PA VA   PA NA SA NO

4   2   5   3   2   5   4   5
MAN GA LA NAM CA SA VVE SIM

3   3   4   1 4 1   4   2   5
PA DHA MAM HA V AI MAN GA LAM

THE FOUR LEVELS OF SOUND

There are countless sounds vibrating in the universe and
within us, at different levels of vibration and subtlety.
The four stages of sound are as follows:

Vaikhari is the grossest form of sound that we hear in the world around us. It is the level of the spoken word, or the sound of striking two objects together. Here, thought is translated into name and form (language). This sound appears from throat to mouth.

Madhyama means "in the middle". It is the transition stage between heard sound and subtle inner vibration. This sound appears from the heart to the throat and is also referred to as unstruck sound.

Pashyanti is sound in the form of light. It can be seen as light but it is a very subtle inner sound and can be heard only by an awakened practitioner. This sound appears from the navel to heart.

Para is the stage of transcendental sound. It is beyond all names and forms. This is the starting point of subtle sound (nadam) which corresponds to the kundalini energy. It is sound which has an infinite wave-length. It is the base and root of sound. Para means beyond time, space and nature. Nothing can affect it. To understand the reality, totality and secret mystical power of sound, para sound must be realized. Only enlightened persons can perceive this sound. The root of this sound is the base of the spine to the root of the navel.

It is difficult to understand the system of para sound. The root of the navel is a great computer. All sound has its source there, then it appears successively in the heart (pashyanti), in the throat (madhyama), and finally from the mouth (vaikhari).

When the kundalini (concentrated pranic energy) moves upward, as it passes different centers, various sounds can be heard. This is called anahat nadam or unstruck sound. When the sushumna path is clear, the kundalini will pierce the *brahmagranthi* (psychic knot in the muladhar), the *vishnugranthi* (knot in the heart chakra) and the *rudragranthi* (knot in the third eye center). Upon reaching the sahasrar, sound becomes soundless. this is the state of super-consciousness, the state where para sound is realized.

When you perceive how life force moves, how it turns into sound and how this sound emanates from its source, the root of the navel, then you will be enlightened.

As human beings our five senses are very limited. Members of the animal kingdom far surpass our powers. For example, we can hear a bell ringing, and if we remain close to the bell after the sound has diminished, we may perceive some vibration. However, long after the continuing sound is imperceptible to us, it can still be heard by a mouse. As a matter of fact, a popular nonviolent method of pest control employs the use of ultrasonic frequencies, which so disturb the sensitive ears of rodents that they flee. Dogs have a very keen sense of smell. Owls and other birds of prey have extremely sharp eyesight. A dolphin can receive the vibration of a call sent by another dolphin over a distance of 1300 kilometers (ca, 788 miles) underwater. This is due to their acute sense of hearing. These animals have developed their subtle "yogic" senses.

Each of the five senses can be classified into different states of perception. As humans we have not developed our subtle powers. Our awareness is so limited, we are like persons sitting in front of a television – we can see and hear the forms and sounds on the screen, but it is impossible for our other senses to be involved in the scene. A yogi on the other hand, has awakened all his subtle senses and can perceive true reality.

Gross sound is incomplete; we must reach its source, the current of *nadam*. If we can approach the science of the Namokar Mantra from all four levels of sound, then the highest result will be attained. Starting with vaikhari repetition, the mantra can awaken all our sleeping powers, bring us to the state of deep meditation, and ultimately to the realization of para sound.

Shanmukhi Mudra

Anahat Nadam may be heard naturally by advanced meditators, or with the help of Shanmukhi Mudra.

TECHNIQUE: Using the fingers of both hands, press the ears with thumbs, eyes with index fingers, nostrils with middle fingers and mouth with ring fingers. (Place little fingers near the chin.) The main pressure is on the ears and eyes. You can experiment with pressure on the eyes until light can be seen – either by direct pressure on the eyeball or inward and upward toward the third eye.

To breathe, release the right middle finger from the right nostril and inhale. Close the nostril again and hold the breath as long as possible without straining. Then exhale through the mouth making a noise like the wind. Do this three times and rest. Breathe normally with eyes closed. Then do another round of three. Do not do more than this since it creates much heat and is very stimulating.

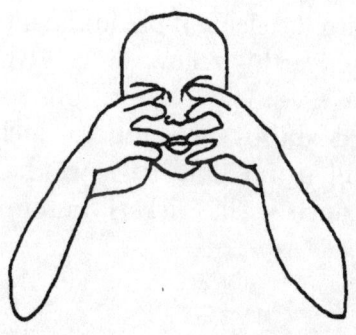

COMMENTS AND VARIATIONS: With the inhalation through the right nostril (pingala – sun side), much heat is created which stimulates the kundalini and allows for more ease in seeing inner light. However, in times of high blood pressure, anger or any "hot" state, the following variation will be more calming and peaceful: Using the fingers to block the senses, just as described above, simply inhale through both nostrils by releasing the middle fingers. Hold the breath and close the nostrils, then exhale through both nostrils by releasing the middle fingers again. This may be practiced ten to twenty minutes.

Shanmukhi Mudra is very useful in preparing us for meditation. We are usually involved with the gross senses; our focus is on the outer world. Shanmukhi Mudra helps us to tune into the inner world by awakening the inner senses. *Shan* means "six"; *mukhi* means "mouth" (door); *mudra* means "pose". We have six doors to the outer world – two ears, two eyes, one nose and one mouth. If we close these six, then we have control of our senses. We can see inner light and hear inner sound – which makes up our essence.

This is normally practiced before meditation. When a busy schedule may not permit time for meditation, then occasionally this mudra may replace meditation because of its centering effect.

The various lights or colors seen during the mudra correspond with the element predominant in you at that particular time (earth – yellow; water – white; fire – red; air – blue; space – smoky). the various sounds (anahat nadam) heard during the mudra include: thunder, ocean, waterfall, drum, flute, bells, wind, cymbal, string instrument, conch shell, crickets, humming bees, and others as well.

PART II
GENERAL MANTRA KNOWLEDGE

THE GURU IS ONE who has harnessed the power of sound. When he gives the disciple a mantra, this sound is alive. It is the same thing as eating. Before we eat, we and the food are separate. But after we chew, digest and assimilate the food, it becomes a part of us. We do not call it food anymore, because it has become a part of our blood, body and life force. It is no longer separate. So, when the guru gives a mantra, the disciple at first is separate from the mantra. But after he or she has spent some time repeating it with love and faith, following the proper techniques, the disciple will completely absorb the mantra. It will be a part of the life force and will work positively for him or her.

It is the duty of the disciple to practice faithfully and follow the guidance of the guru. The disciple should not only have the attitude that the guru will give all the benefits. He will indeed do so, but first the disciple must forget who is the giver and who is the receiver. The mantra will give help only if the disciple can dissolve the self into the mantra. It is the guru's duty to give the mantra energy, so that the disciple will easily be able to feel the current which the mantra creates. By the power of the guru, the mantra works in the disciple, and eventually the disciple and the mantra become one.

PROPER PRONUNCIATION

Basically, it is always important to repeat the mantra with correct pronunciation. There are special times such as sunrise, noon, sunset, midnight and auspicious times, such as the Indian festival of Diwali, when correct pronunciation is more important. These are powerful times which can bring the fulfillment of the desired purpose of the mantra.

If we do not use a machine properly, it will not work. It is the same with sound. It must be repeated with faith, no doubt, but proper pronunciation whether mentally, softly or loudly is necessary. This will bring a more beneficial result.

VEGETARIANISM

It is necessary to eat pure food in order to repeat a mantra. By eating meat, we are involved in the killing of animals. By killing their life force, we disturb our own. Then our etheric body cannot become pure and strong. The Namokar Mantra is based on non-violence. A pure and natural vegetarian diet is necessary for spiritual progress and the success of mantra repetition.

True, there is a certain amount of violence in killing plants for food as well as animals, but animals possess a greater number of senses. An animal feels, sees, smells, tastes and hears and therefore, experiences fear and much more intensified emotion than a plant, which only has the sense of touch and sound. It has been proven in studies that plants are sensitive to certain sound vibrations, for example, they grow much better when classical music or chants are played in their vicinity but don't do well with rock music or jazz. We can compare the different vibrations of a slaughterhouse to a vegetable garden or market.

Black tantric mantras call for the exact opposite system of what we are following. They involve techniques requiring violence, meat and alcohol consumption, and sex.

MANTRA AND BREATH

Breath is prana or life-force. Sound produces pranic energy. When chanting *Hreem,* for example, we can see how sound actually produces breath. *H* is the seed of space, *R* and *E* are seeds of fire. These sounds create oxygen, carbon dioxide and all types of life force, affecting the mechanism of the entire body. If we repeat *Hreem* quietly, moving only the tongue and keeping the lips apart, it is possible to continue for nearly an hour without taking a breath. This is the automatic breath that comes with mantra.

Also, if we consciously control the number of breaths we take when repeating mantra, we can get a greater effect. The mantra *shastras* or commentaries explain that only one breath should be used for a mantra – at the most, three breaths. It is most powerful when the entire mala is repeated in one breath. With control of the breath, mantra, prana and the mind merge. This harmony comes when the senses and the intellect are at rest.

MALAS AND HASTANGOLI

When using a mala (a rosary), count with the thumb and middle finger or ring finger of the right hand (not the index or little finger). The mala should not touch the floor. The current produced from the mantra can be lost if it touches the floor. Hold the mala up and repeat the mantra mentally or aloud. For more energy, you may hold the mala at the navel or solar plexus in the morning, at the heart in the afternoon, and at the third eye in the evening. (The last is difficult and is not necessary unless specifically suggested for you.)

A mala generally has 108 beads (totalling the cosmic number 9 when the three digits are added together). This number is suitable for all purposes. We also use malas of 54 or 27 beads. Other numbers may also be used for specific purposes.

Without a mala, a beautiful system of japa uses visualization in the lotus of the heart. Visualize the eight-petalled lotus in the heart center. Imagine twelve bright golden dots in each petal and twelve dots in the center of the lotus. Repeat the mantra for each dot. This adds up to 108 times. Hastangoli is another method of japa using the fingers of the hand to count. The diagram on the following page includes the general techniques of counting on the fingers. However, there are many systems of hastangoli for various purposes. Using your right hand, point with the tip of the right thumb to the phalanges of the fingers following the numbered system. Three different methods are presented on the following page.

Malas are made of gemstones, metal, seeds or wood which are directly related to the rays of the planets and their corresponding colors. The guru can recommend the proper mala for a particular mental or physical condition, or for a particular mantra, considering the astrological aspects, color science and the objective of the japa.

Malas and Their Effects

GOLD Sun energy

SILVER Moon energy; emotions, alleviates restlessness, peace, love

RUBY Sun energy; navel, circulation, digestion, sushumna

PEARL Moon; emotions, peace, harmony

CORAL Mars; purification of mind and blood, wealth, health

EMERALD Mercury; fixed thoughts, wisdom, strength, boldness, life force

DIAMOND Venus; general beauty, worldly attainments, wealth, health and love

SAPPHIRE Saturn; wisdom, intellect, life force

ONYX North and south nodes; removal of bad luck and karmic problems

CAT'S EYE North and south nodes; protection from evil eye

AMETHYST Saturn; good for right eye trouble, use instead of sapphire as it gives the same effects

AMBER Jupiter; stomach, liver, kidneys, disinfectant

TURQUOISE Saturn; prana, removal of fears

MOONSTONE Moon; can be used instead of pearl as it gives the same effects

JADE Mercury; heart, respiration, harmony, prana

TOPAZ Jupiter; wisdom, discipline, sushumna

SANDALWOOD & TULSI General good purposes

RUDRAKSHA Sun and Saturn; worldly or spiritual benefits

LAPIS LAZULI Saturn; healing, circulation, skin, eye problems, depression

QUARTZ or CRYSTAL Moon and Venus; emotions, peace, intellect, love

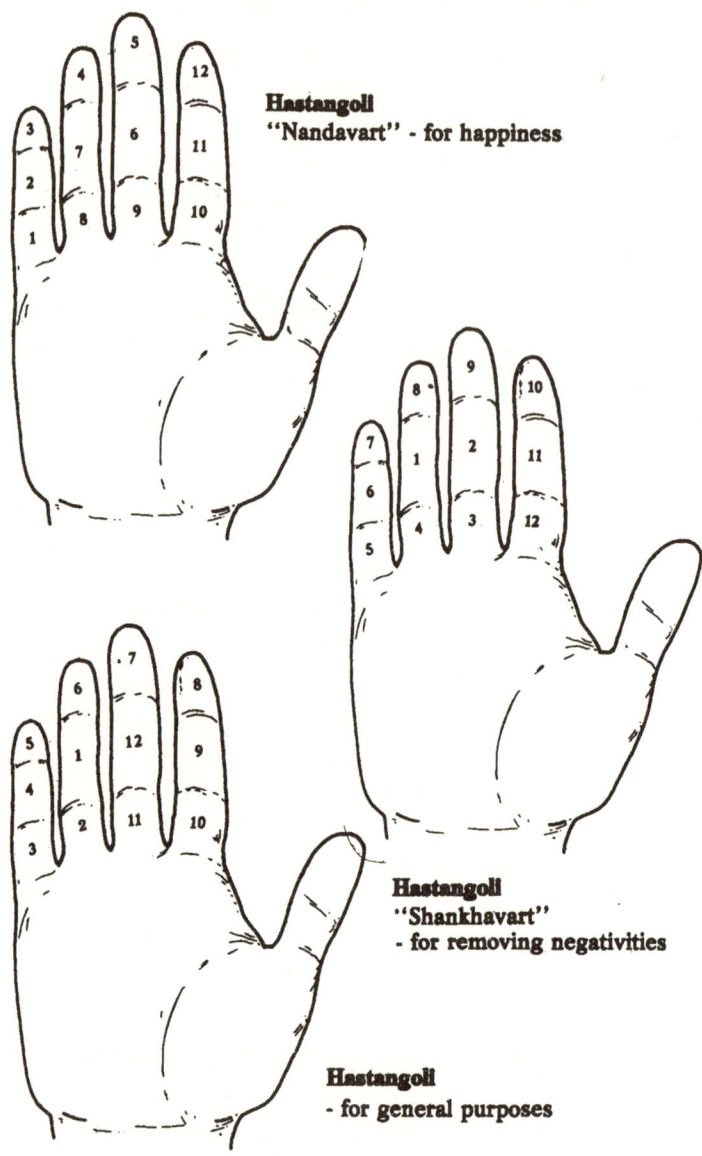

Hastangoli
"Nandavart" - for happiness

Hastangoli
"Shankhavart"
- for removing negativities

Hastangoli
- for general purposes

Techniques shown have been taken from Mahaprabhavik
Navasmarna *by Sarabhai Navab.*

ASANAS AND MUDRAS

Mantra can be repeated mentally anywhere at any time. However, it should not be repeated aloud in an unclean place, or if the body is not clean.

One can repeat mantra sitting, standing, walking, running or lying down. but, in order to get the full benefit or to attain a desired objective, the mantra must be repeated in a proper seated or standing posture with the back erect, using visualization.

Padmasana or Siddhasana* are ideal postures for mantra repetition, while holding the mala with the right hand at the heart level. Mantra repetition without the use of a mala is recommended in the following positions:

1 JINA MUDRA – stand erect with feet slightly apart. Arms straight with hands down at sides and fingers straight. Close eyes half-way.

2 YOGA MUDRA – sit in Padmasana with right hand on top of left at navel, palms upward.

3 GYAN MUDRA – sit in Padmasana with the left hand on the left knee, thumb and little finger touching, palm of hand facing upward and three first fingers are straight. Right arm is held straight up with all straight fingers pointing upward.

Different colors of seats used during mantra repetition will give various effects. Normally, white, yellow, orange

---

* Padmasana – lotus posture, where right leg is bent and right foot is placed at the root of the left thigh. Then the left leg is bent and the left foot is placed over the right, with the soles of both feet turned upward.

* Siddhasana – bend left leg and place left heel near perineum. then bend right leg and place foot over left ankle, with heel against pubic bone.

or red are suggested. White is for peace and purification. Yellow or orange is for wisdom and happiness. Red is for stimulation and worldly success. Green or blue interfere with the warmer colors at the base of the spine when seated, therefore, they should not be used.

CONTACTING DIVINE POWERS

The yogic and devotional systems of prayer are different. For example, in the devotional system when we repeat the Saraswati Mantra, we are calling the Goddess of Wisdom to appear to us. We repeat the mantra over and over again, and it establishes its trace in our memory. When that trace becomes more powerful, then her form actually appears by our will. The faithful repetition awakens the form and power of the mantra.

But in the yogic system of prayer, the attitude is different. We understand that Saraswati represents a main nadi ending at the tip of the tongue. This center is the center of wisdom, and controls the power of speech or sound, and the power of taste. With repetition of the Saraswati Mantra, wisdom, intellect and supernatural powers (clairvoyance, ESP – extra-sensory perception, *etc*) will awaken. The power of taste will also develop, making it possible to taste a food prepared even a thousand miles away.

It is important to realize that the deities actually represent various powers. The mind is the creator of any manifestations which appear. The mind has infinite power. The Arihantas have taught that we have infinite knowledge, perception, conduct and energy. The power is ours – any power – physical, psychic (mental) spiritual. It only needs to be awakened by sound vibration.

PART III
TECHNIQUES FOR PRACTICE

THE FOLLOWING TECHNIQUE will awaken the Namokar Mantra within us:

On a piece of paper write the words Namo Arihantanam in Sanskrit letters.* Stare at these words for three minutes or until the eyes can no longer remain open without blinking. Then close the eyes and visualize the letter Na [न]. Continue letter by letter, until Namo Arihantanam is completely visualized. Practice this concentration twenty minutes per day. When the words appear at will, in their proper color, without the aid of the paper, this shows that the mantra is awakening. Eventually the entire mantra can be visualized. If you can concentrate for two hours on even one line, the Jinabimbam, which is the image in white light of the Arihant, will appear and the letters will disclose all secrets of power and knowledge.

For the beginner, it is sufficient to concentrate only on the latter Na [न] for a few days. First, write the letter on a piece of paper and follow the concentration practice as described above. Then close the eyes and visualize only the letter Na for twenty minutes. If the Sanskrit letters are very difficult to visualize, you may use Roman letters.

When visualizing one or more of the letters of the Namokar Mantra, imagine that you are writing the letters in space in a large size, almost the size of your body.

There are special times when the individual lines of the Namokar Mantra may be visualized to strengthen the color rays more effectively:

---

* See page 17 for Namokar Mantra in Sanskrit letters.

Namo Arihantanam	white	full moon
Namo Siddhanam	red	sunrise
Namo Airiyanam	yellow	noon
Namo Uvajjhayanam	blue	sunset
Namo Loe Savva Sahunam	black	dark of night

Following special methods for awakening mantras will speed up the desired results of knowledge, supernatural powers or spiritual advancement. Generally, a specific number of repetitions should be accomplished over a three-day period. Fasting or partial fasting should also be observed at that time. Specific colors of clothes, asanas, mudras, malas and dates (time) depend on the particular mantra, objective and astrological aspects.

PRACTICAL ADVICE FOR REPETITION

We should prepare ourselves to chant the Namokar Mantra by feeling love, devotion and oneness. Visualize the Arihant and become the Arihant. Purify the mind of desires, thoughts and negativities. The body must be clean and peaceful and the seat should be clean. Sit comfortably with the back erect, facing north or east.

Apply the Namokar Mantra to your body with colors* flve times, then repeat one mala of the mantra (108 times). The mantra will nor become completely active unless it is first applied in this way before repetition. For more complete preparation apply the mantra first to your fingers with color (and then to your body). Apply it to the fingers of both hands at the same time, one pair of fingers at a time, starting with the thumbs:

Namo Arihantanam	thumb	white
Namo Siddhanam	index finger	red
Namo Airiyanam	middle finger	yellow
Namo Uvajjhayanam	ring finger	blue
Namo Loe Savva Sahunam	little finger	black

This advice is for your daily morning japa or special times of repetition. Of course, you can use the Namokar Mantra anytime, anywhere for help and protection. Once you establish a relationship with the mantra, even though you may sometimes forget it, it will never forget you.

These are the three main systems for repetition of the Namokar Mantra or any mantra:

mentally	will bring the power to predict the future and develop the power of speech
mentally with visualization	will remove inner negativities, bring spiritual progress and self-realization
vocally (softly)	will bring worldly achievements
vocally (loudly)	will improve the atmosphere of the environment, attract angels and divine powers, will destroy poverty

CONTROLLING THE FIVE AIRS[1]

The technique for controlling the five airs by the Namokar Mantra is as follows:

Begin with these general steps—

1 Sit in Siddhasana

2 Exhale and hold Mulbandha (base lock)[2] and Uddiyan Bandha (stomach lock)[3] inhaling when necessary

3 Shanmukhi Mudra[4]

4 Inhale with Kaki Mudra (purse lips and partially block opening of mouth with tongue; this will make a noise when inhaling by mouth. Exhale through nostrils.

5 Hold breath and mix prana and apana. Apana goes up from the anus. Prana goes down from the heart. They mix by Uddiyan Bandha. This is called Shaktichalini Mudra.

Then, to control prana—

For five minutes repeat *"Om Hreem Yeim Namo Arihantanam"* and visualize the seed sound of yeim [यैं] in blue or green at the heart.

To control apana—

Repeat *"Om Kleem Peim Namo Siddhanam",* and visualize peim [पैं] in black at the base of the spine.

---

1 *See pages 34 to 37 for a discussion of the five airs and their functions.*

2 *Contract the anal, perineal and genital area. This area is lifted up toward the spine.*

3 *Contract the abdominal muscles. The diaphragm is lifted and the abdominal organs are pulled back toward the spine.*

4 *See page 63 for details.*

Shree Siddha Chakra

Shree Rishi Mandal Stotra

To control samana—

Repeat *"Om Kshoum Veim Namo Airiyanam"*, and visualize veim [वैं] in white at the navel.

To control udana—

Repeat *"Om Shreem Roum Namo Uvajjhayanam"*, and visualize roum [रौं] in red at the soft palate (near uvula).

To control vyana—

Repeat *"Om Bloom Loum Namo Loe Savva Sahunam"*, and visualize vyana [व्यों] in mixed colors at the throat. For controlling vyana include Jalandhar Bandha (chin lock)[5] but omit Shanmukhi Mudra.

SPECIAL VARIATIONS OF THE NAMOKAR MANTRA

Variations for Great Protection

The Namokar Mantra and the Powers from the Six Directions:

Om Hreem Shreem Namo Arihantanam
Om Hreem Shreem Namo Siddhanam
Om Hreem Shreem Namo Airiyanam
Om Hreem Shreem Namo Uvajjhayanam
Om Hreem Shreem Namo Loe Savva Sahunam
Om Hreem Shreem Namo Nanassa
Om Hreem Shreem Namo Dansanassa
Om Hreem Shreem Namo Charittassa
Om Hreem Shreem Namo Tavassa

---

5 *Contract the muscles of the neck and throat to bring the head forward with the chin touching the jugular notch between the collar bones.*

This mantra is repeated five times while you stand and face each direction, beginning with east and then south, west and north. Lastly, while still facing north, turn the face downward, and finally turn the face upward. With this prayer, we invoke the following powers of the Six Directions:

We call on the powers of the east—this is the side of the rising sun, the side of the Tirthankaras, for knowledge, perception, and an unfolding of consciousness and liberation.

We call on the powers of the south—for the fulfillment of worldly desires and wealth.

We call on the powers of the west—this is the side of iron, the side of machinery, for worldly knowledge, protection from accidents and all troubles.

We call on the powers of the north—this is the side of monks, saints and religious practitioners, for harmony, peace and devotion.

We call on the powers from below—for opening all secrets of the earth, secrets of wealth and precious things, contact with the souls of the Arihantas.

We call on the powers from above—this is space, the ocean of energy, for all good energies (which enter the body from above through the top of the head), for brotherhood, love, sound vibration, health, contact with the Siddhas and astral beings.

Energy is working and moving everywhere. but it is positive energy that we need – the divine current to awaken the higher mind. We call on good energy from the four directions, and from below and above, for purification, protection and spiritual awakening.

Divine vibrations from the Arihants come from each direction. Below in the hollow earth are souls of the

Arihantas who have not yet incarnated. Extraordinary, powerful souls are everywhere. We contact these great souls and call on their positive energies to remove the negative, destructive forces which cause obstacles, enmity, accidents and catastrophes.

By this mantra we call on the powers of *Om, Hreem, Shreem,* Arihant, Siddha, Acharya, Upadhyaya, Sadhu, Right Knowledge, Faith and Conduct (*Nanassa, Dansanassa, Charittassa*). We call on them for protection, peace, knowledge, health, worldly attainments, total bliss and liberation.

After we have reached a higher level of consciousness, then we can lead others to attain the same. On a large scale, if the world should be threatened with catastrophes and destruction, then great protective powers can be called upon to save mankind.

This is the unique quality of this prayer – maintaining the attitude of aspiring not only to personal·protection and attainments, but ultimately of reaching out to all with brotherhood and oneness.

Astral Projection and the Namokar Mantra

First mentally ask permission from the guru, which also means your inner guide or higher self, to practice this technique. Bow to Lord Adinath, the first Tirthankar, and Chakreshwari, the goddess of the chakras.

Preparation—

1 Repeat the following three times:

Om Namo Arihantanam, Namo Siddhanam
Namo Agasgaminam,[1] *Om Namah*

---

1 *Agasgaminam is the lord of astral travellers.*

2 Apply the following lines of the Namokar Mantra to the body by using the following visualization:

Om Arihantanam Namah — heart
Om Siddhanam Namah — head
Om Airiyanam Namah — transpersonal point
Om Uvajjhayanam Namah — imagine an armor
    of blue light on the entire body
Om Loe Savva Sahunam Namah — imagine
    holding a big weapon

3 Visualize the Arihant in the heart center. State that you want to travel astrally and ask for protection.

4 Repeat the mantra in step one 108 times, then follow the technique for the actual projection.

Technique—

1 Ask permission of the guru for help and protection.

2 Stand with hands raised up; when you feel a tingle of energy in them, put them down.

3 Perform Bhastrika (bellows breathing).[2]

4 Lie or sit down and close your eyes.

5 Chant Om or Arhum several times, long and peacefully.

6 Give up all conceptions of the self and ask "who am I?"

7 Relax all parts of the body.

8 Take long quiet breaths.

9 Concentrate on the solar plexus.

---

2 *Exhale forcefully through the nose making noise. Inhale forcefully, also making noise. Continue rapidly in this fashion, balancing the length and force of inhalation with that of exhalation. Do this ten, twelve or twenty times. Then breathe deeply and slowly two or three times.*

10 Visualize the ejection of a silver-grey mist of astral material from the solar plexus, which is attached at this center by a silver cord.

11 Visualize that this small silver-grey cloud is taking the form of your body.

12 Transfer your consciousness, the seeing, hearing and feeling powers to the astral body.

13 Direct this astral body where to go, how long you wish it to stay, and when you want the body to return.

14 Draw the astral body back through the silver cord by long deep breaths. It enters into the physical body through the solar plexus. Make sure it returns completely "in place".

From three to five in the morning is the best time for astral projection on full-moon days. Three days before and three days after the full moon are also favorable. None-theless, the full moon day is the most conducive time.

Balancing Planetary Effects

By using the following variations we can balance the effects of the planets, as understood astrologically:

SUN	*Om Namo Siddhanam*
MOON	*Om Namo Arihantanam*
MARS	*Om Hreem Namo Siddhanam*
VENUS	*Om Hreem Namo Arihantanam*
MERCURY	*Om Hreem Namo Uvajjhayanam*
JUPITER	*Om Hreem Namo Airiyanam*
SATURN	*Om Hreem Namo Loe Savva Sahunam*
SOUTH & NORTH NODES	*Om Hreem Namo Loe Savva Sahunam*

More Variations

The following variations will create their corresponding powers:

ARRESTING POWER (the ability to stop something)
Om Namo Arihantanam Thah Thah

DISTURBING POWER
Om Namo Arihantanam Fat Swahah

MESMERIZING POWER
Om Namo Arihantanam Namah Swahah

ATTRACTING POWER
Om Namo Arihantanam Voushat

SUBMISSION
Om Namo Arihantanam Vashat

REMOVES ENMITY
Om Namo Arihantanam Hoom

KILLING (destroys negative power in others)
Om Namo Arihantanam Ghe Ghe

PEACE
Om Namo Arihantanam Swahah

STRENGTH
Om Namo Arihantanam Swadha

WEALTH
Om A, Si, Ā, U Saya Namah

PEACE (removes enmity)
Om Arhum A, Si, Ā, U Saya Namah

HEALTH
Om Arhum A, Si, Ā, U Saya Namo Arihantanam Namah

FULFILLS DESIRES
Om Hreem Shreem Arhum A, Si, Ā, U Saya Namah

TO COMPLETE UNFINISHED WORK (repeat 1225 times)
Om Hram Hreem Hroom Hroum Hrah A, Si, Ā, U Saya Namah

FOR ALL KINDS OF ACHIEVEMENTS (repeat 12 malas)
Om A, Si, Ā, U Sa Choo Loo, Choo Loo, Hoo Loo, Hoo Loo, Koo Loo, Koo Loo, Moo Loo, Moo Loo, Icchiyam Me Koo Roo, Koo Roo, Swahah

FOR RAIN (repeat 1000 malas*)
*Om Eim Hreem Shreem Kleem
Om A, Si, Ā, U Saya Namah*

TO REMOVE GHOSTS (repeat 42 times)
Om Hram A, Si, Ā, U Sa Pretadikan Nashaya Nashaya

TO REMOVE POISON (repeat in one breath only)
Om Hreem Arhum A, Si, Ā, U Sa Kleem Namah

FOR VICTORY IN A COURT CASE
*Om Hamsah Om Hreem Arhum
Eim Shreem A, Si, Ā, U Sa Namah*

TO STOP FIRE
*Om Namo Om Arhum A, Si, Ā, U Sa
Namo Arihantanam Namah*

---

* *the Guru will give the secrets of the technique for performance of the mantra*

The variations above have been taken from Namaskar Mantra Siddhi *by Shri Dheeraj Lal Tokarsi Shah.*

Mantra for Victory and Fame*

Technique—

Repeat the following once upon beginning repetition of the mantra below for the day:

Om Hreem Arhum Namo Uvajjhanam

Then repeat the following mantra for twenty-one days, ten malas every day:

Om Hreem Shreem Hreem Krom Vashat Swahah

During japa you must sit on a yellow cloth, use a yellow mala, burn sandalwood incense and face south. Visualize the mantra (as shown on page 99) in the third eye center for more power.

Effect—

This mantra will bring fame and victory or success in endeavors.

---

* *This technique has been taken from* Mahaprabhavik Navasmarna *by Sarabhai Navab. As with all the techniques, its use should never violate the principle of non-violence and the feelings of fellowship and oneness with other beings.*

VISUALIZATION OF MANTRA
FOR VICTORY AND FAME

MEDITATION PRACTICES

Meditation on Arhum

Technique—

First, through your power of imagination establish *Arhum* (in Sanskrit letters: see page 98) in a golden color in the eight-petal lotus of the navel center. Visualize and concentrate on *Arhum* there. Then it will become activated (use imagination) and move out of the body and into space. Visualize *Arhum* in space and see it pure and shining like moonlight. Again *Arhum* begins to move and from space it will enter into the mouth, cross the third eye center, forehead, fontanel and palate, showering nectar and emanating white light throughout the chakras and the body. You can begin the practice with normal breathing, then hold the inhalations as you visualize *Arhum* distributing nectar and light through the chakras.

Effect—

This technique will awaken all the powers of *Arhum*. It will remove sadness and depression and give stable happiness. Concentrating on the *H* [ह] of *Arhum* will bring the fulfillment of desires. Merging with *Arhum* will give completeness and perfection.

Techniques have been taken from Mahaprabhavik Navasmarna *by Sarabhai Navab.*

Meditation on Letters to Remove the Three Knots*
Technique—

Begin by visualizing *Arhum* in white in the center of the eight-petalled lotus of the navel. In the circle around the center is *Hreem* (eight times), and in the circle around that are all the vowels of the Sanskrit alphabet. Then, repeated in one of the petals of the lotus, are the vowels again. And in the remaining seven petals the consonants and semi-vowels appear.

Effect—

This is a difficult practice for those not familiar with the Sanskrit alphabet. however, the result is tremendous. To make it simpler, we have presented it in Roman letters.

By this practice the three *granthis* (knots) can be broken. These knots block the free flow of pranic current through the sushumna, thus preventing the rise of the kundalini. The powers of the vowels (governed by the moon) will pierce the knot at the root center and awaken knowledge; the powers of the consonants (governed by the sun) will pierce the knot at the heart center and awaken our powers; and the powers of the semi-vowels (governed by fire) will pierce the knot at the third-eye center and awaken perception. Ultimately, the self merges into *Arhum;* the individual "I" merges into the principal "I".

Also, by concentrating on the letters in this manner, our ability to learn any language is greatly increased.

---

* *See pages 98–102 for the Sanskrit alphabet presented in Roman script.*

Meditation on A, Si, Ā, U, Sa

Technique—

Visualize the following letters in their corresponding centers:

A   navel
Si   top of the head
Ā   mouth
U   heart
Sa   throat

Effect—

This practice is for betterment and total salvation. it can be visualized by concentrating on one letter individually or by concentrating on all five. These five letters awaken the powers of the five divine personalities of the Namokar Mantra:

A – Arihant	purification, protection, perfection
Si – Siddha	perfection, total enlightenment, energy
Ā – Acharya	will power, discipline, knowledge
U – Upadhyaya	knowledge, creativity, balance of body, mind and soul
Sa – Sadhu	power to remove negativities and attachments

Meditation on *Om Namo Arihantanam*

Technique—

First, concentrate and visualize the beautiful, shining, white figure of the Arihant. Then visualize a lotus of eight petals in the navel or other centers.

In Sanskrit or Roman letters visualize *Om* in the center of the lotus, then with one sound in each petal place *Om* (again), NA, MO A, RI, HAN, TA, NAM.

(This system can also be used with *Om Namo Siddhanam, Om Namo Airiyanam, Om Namo Uvajjhayanam* and *Om Namo Loe Savva Sahunam.* The number of petals will vary according to the number of sounds.*)

Effect—

When you apply this practice you can awaken the four powers of the Arihantas:

1. Knowledge
2. Sound and the spoken word
3. Honor and respect
4. Power to remove obstacles

You can choose which one of the five centers below you wish to concentrate on. Each one will give a different effect:

fontanel – to remove disease
third eye – to awaken intuition
throat – to awaken pure knowledge
heart – to awaken perception
navel – to awaken psychic powers

---

* *See page 60 for proper separation of sounds in Namokar Mantra.*

MEDITATION ON NAMO ARIHANTANAM

SANSKRIT LETTERS

Vowels

अ a – unity of soul, base of *Om* sound, eternal, omnipresent, pure, enlightened, power

आ ā – creator of intellect, eternity, power, wisdom, fame, wealth, fulfils desires, force of attraction, working universally

इ i – creatorof fire seed, wealth, action, easy work, peaceful power of nourishment

ई ī – creator of nectar seed, giver of knowledge, fulfils work, attracting, arresting, making shivering in body, pure, power of speech

उ u – miracle powers, disturbs mind, easily bored, can cause death, the essence of all powers

ऊ ū – root of attractive seeds, destructive power, the unbearable power of expulsion

ऋ ṛi – seed of fire, producer of all main seeds, root of goodness, perfection, secret wealth of mind and soul, fulfillment, distrubting agitation of mind

ऌ lṛi – announcer of truth, destroyer of speaking power, producer of seed of wealth, medium of self-realization, hypnotic influence creating hatred

ए e – destroys obstacles, perfect, active, giver of strength, the purest

ऐ ai – water seed, producer of positive electric current, strengthening, attractive, helps growth, fulfils difficult work, calls angel powers, the purest power of attracting all persons

ओ oo – producer of *Hreem* sound, helper of all sounds ending in m, the ever pure word God, giver of wealth, helper of hard work, destroyer of karma

औ au – killer, creates detachment, producer of many seeds, attractive, unconcerned, quick fulfiller of work

अं aṁ – seed of space, independently has no power, chief destroyer of karma, messenger of black hole, sponsor of soft powers, base of wealth seeds, hypnotic influence on animals

अः ah – main seed among peace seeds, no independent power, destroyer of death

Consonants

क ka – seed of power, impressive, producer of joy and comfort, helper for fertility and sex energy, antidote to all poisons

ख kha – seed of space, helper to fulfil desires and work, creates confusion in friendships, powerful to change the minds of others, disturbing

ग ga – opposes disharmony, greatest remover of all obstacles

घ gha – arresting seed, destroyer of obstacles, hypnotizer, killer, giver of stability

ङ gña – destroys enemies, producer of destructive seeds, antidote for all poisons

च ca – creates deficiencies, helpful with other seeds, separation, powerful to change the mind, the cruel, destroyer of amenities

छ cha – creates bondage, destroys power, easy work, water seed, helper of Hreem, indication of shadow, greatest remover of evil spirits

ज ja – helper of innovative work, attractive, powerful healing powers, unrivalled destroyer of black magic, emotional

झ jha – same as *ja*, producer of wealth

ञ ña – arresting, attracting, conqueror of death, against meditation

ट ṭa – seed of fire, increases work of fire, destruction, curer of disease, the fortunate

ठ ṭha –unlucky, destroyer of peace and easy work, helper of hard and painful work, creates weeping and tension, fire seed

ड ḍa – explosive, makes angel powers appear, useful for simple work, antidote for poisons, combined with other seeds will produce five elements

ढ ḍha – immovable, seed of killing, destroyer of peace, gives power and strength, giver of fortune

ण ṇa – bestows peace, seed of space, explosive power, giver of psychic powers

त ta – seed of attraction, inventor of powers, fulfils all work and desires with help of *Eim*, joyful

थ tha – destroys obstacles, helper of seed of wealth, attractive with vowels, pure, producer of virtues

द da – destroys karma, creates appearance of power of soul, helper of seeds of attraction, giver of greatness

ध dha – helper of *Kleem* and *Shreem*, curer of deadly fevers

न na – power of soul, producer of water element, provides easy work, controller of self, helper of realization, peaceful, giver of enjoyment and liberation

प pa – helper of realization, power of water element, helper of all kinds of work, grave, destroyer of poisons

फ pha – fulfiller of difficult and important work, power of water and air elements, remover of obstacles, giver of psychic powers

ब ba – remover of obstacles, good for perfection, destroyer of bad habits, good-looking

भ bha – energy for killing and injuring, disturbing, creates obstacles in divine practice, enemy of seed of wealth, inhibits pure work, destroyer of evil spirits

म ma – creator of perfection including perfection in spirituality and materialism, help to fulfil desire to bear children, hypnotizer of opponents

Semi-vowels

य ya – peacemaker, useful for important work, helper in making friends and attaining positive things, sanctifier

र ra – seed of fire, fulfils work, producer of all main seeds, powerful, eternal

ल la – seed of wealth, closest friend of *Shreem*, goodness, protector of world

व va – giver of perfection, creates miracles with *h* and *r*, seed of Saraswati – wisdom, remover of ghosts and disease, remover of obstacles, arresting power, fulfils worldly desires, pure, water

श śa – peaceful, giver of virtue, no outstanding power, white

ष ṣa – producer of seeds for calling, giver of perfection in worldly matters, arrests fire and water, extra-ordinary work with other sounds, fulfils dangerous work, fearful

स sa – fulfiller of desired work, can use with all seeds, for peace and strength most useful, destroys karma covering knowledge and perception, producer and helper of *Kleem*, helper of self-realization and enlightenment

ह ha – producer of peace, strength and good work, most useful for practice, free, helper of wealth, fulfils desire for children, pure, seed of space, producer of all seeds, destroyer of karma

PRONUNCIATION GUIDE

The following is a close approximation of the pronunciation of Sanskrit letters which have been transliterated into English. (The key below does not include all letters.)

Vowels

a –   pronounce as vowel sound between 'u' in "n*u*t"
      and 'e' in "n*e*t"

ā –   "*a*h"

i –   "s*i*t"

ī –   "k*ee*n"

u –   "l*oo*k"

ū –   "b*oo*t"

ṛi – slightly trilled sound of 'ri' less pronounced than
      that of Italian or Spanish

e –   "l*a*ke"

ai – "l*ie*"

o –   "c*oa*t"

au – "*ow*l"

NOTE: Arhum *and* seed sounds such as hum, lum *and* vum *as found in this text are pronounced with the a [अ] vowel sound. The letter 'u' has been used in the spelling of these words to discourage the English reader from pronouncing them with a "short a" as found in "hat" or "lap".*

Consonants

ṁ –   "ng" sound as in "ru*ng*" without closing the throat

c –   "*ch*ild"

ñ –   "*gna*" sound in "lasa*gna*"

ṭ,ḍ,ṇ –   these are palatal sounds pronounced with the tongue turned upward and backward touching the palate (a hard sound) – no English equivalents

t,d,n –   these are dental sounds pronounced with the tongue touching just behind the front teeth (a soft sound) – no English equivalents

ś –   "sh*out*"

ṣ –   pronounced as "sh" but with the tongue turned upward and backward as in 't', 'd' and 'n' – no English equivalents

s –   "*s*it"

Aspirates such as "th" and "bh" are pronounced with a slight exhalation of air similar to "t-h" in "boat-house" and "bh" in "abhor".

SUMMATION

AS LIGHT IS TO DARKNESS, the Namokar Mantra illuminates and awakens the divine qualities of the soul. It is not a religious ritualistic prayer, but an eternal universal expression of perfection. The Namokar Mantra holds the science of life within itself. It is the key to the divine treasury of knowledge.

When faith is based on knowledge from experience, it becomes unshakable. By mindful repetition and by following the techniques relating to the Namokar Mantra, you can experience its power first-hand in all aspects of life. My blessings to all of you, that your practice may be one-pointed and steady, and that the river of your faith may run deep. The power of the Namokar Mantra will awaken within you and transform you.

This is the way of true humility and liberation. Our goal must be to become the Namokar Mantra.